Measuring Complexity in Early Bronze Age Greece

Measuring Complexity in Early Bronze Age Greece

The Pig as a Proxy Indicator of Socio-Economic Structures

Melanie A. Fillios

BAR International Series 1722
2007

Published in 2016 by
BAR Publishing, Oxford

BAR International Series 1722

Measuring Complexity in Early Bronze Age Greece

ISBN 978 1 4073 0205 8

BAR Publishing is the trading name of British Archaeological Reports (Oxford) Ltd. British Archaeological Reports was first incorporated in 1974 to publish the BAR Series, International and British. In 1992 Hadrian Books Ltd became part of the BAR group. This volume was originally published by John and Erica Hedges Ltd. in conjunction with British Archaeological Reports (Oxford) Ltd / Hadrian Books Ltd, the Series principal publisher, in 2007. This present volume is published by BAR Publishing, 2016.

Printed in England

BAR
PUBLISHING

BAR titles are available from:

BAR Publishing
122 Banbury Rd, Oxford, OX2 7BP, UK
EMAIL info@barpublishing.com
PHONE +44 (0)1865 310431
FAX +44 (0)1865 316916
www.barpublishing.com

TABLE OF CONTENTS

LIST OF TABLES

LIST OF FIGURES

Acknowledgements

This monograph was originally presented as a doctoral dissertation in the Department of Anthropology at the University of Minnesota. This research would have been far more difficult without the help of so many wonderful people. My deepest thanks go first to my advisor, and friend, Greg Laden, without whose support and patience this work would not have been possible. I also owe a debt of gratitude to the Department of Anthropology and the Graduate School at the University of Minnesota, as well as the Helike Foundation, for the monetary supported they provided for my doctoral dissertation fieldwork in Greece.

Thanks are especially due to all those individuals who made fieldwork and research in Greece possible, among whom are Dr. David Reese, Dr. Martha Wiencke, The American School of Classical Studies at Athens, Anna Toussant, L'Ecole Francais d'Athens, The Archaeological Museum of Argos, The Ephoreia of Nauflio, Dr. Sevi Triantafilou, Dr. Dora Katsonopoulou, Dr. Steven Soter and all the amazing individuals who slogged through the heavy Bronze Age mud. I would also like to thank the additional members of my doctoral committee, Dr. Martha Tappen, Dr. John Sodeberg and Dr. Fred Cooper, for their comments and advice. Special thanks also must be given to Professor Paul Halstead, who graciously provided me with a copy of his unpublished faunal data from Tsoungiza.

Most importantly, thank you to everyone who patiently read, edited, and reread drafts of this research. Your time is deeply appreciated and words fail to adequately express my gratitude. Thanks that cannot be measured go especially to Martin Cooper for his endless help, encouragement and understanding. This research is the not the product of one person's work, but the result of endless support. Nevertheless, any errors are the sole responsibility of the author.

PART I

CHAPTER 1

ANIMAL DOMESTICATION IN THE EARLY GREEK BRONZE AGE

"*A society's cuisine acts as a language through which it unconsciously expresses its structure*." – Claude Lévi-Strauss

"*For the chicken the egg demands involvement, but for the pig bacon demands total commitment*." – John Price

I. Introduction to the Topic

The interpretative potential of faunal analysis has long been underestimated in Greek archaeology. In the absence of written texts, zooarchaeology can be used to address economic organization – a central component of social complexity. Archaeological excavation has revealed variation among Early Bronze Age settlements in Greece. This variability in settlement size, coupled with evidence for craft specialization and possible administrative centers, suggests the naissance of a socially complex society. This examination suggests an alternate approach to complexity in society; it employs faunal analysis to address whether evidence for social complexity exists in the raising of livestock.

Social complexity in ancient societies has been studied by historians, classicists, archaeologists and anthropologists alike. Material remains such as written texts, pottery, personal ornament, architecture, and art have provided the main corpus of evidence examined. Yet the most basic necessity has traditionally been over-looked - food. Subsistence strategies are a central facet in the lives of every individual, poor or wealthy, and are influenced by economic and ecological constraints, and by social mechanisms. Food choices are also a reflection of these and other factors – especially in the social realm. Amazingly, animal bones, often the most prevalent body of material recovered at archaeological sites, have not figured prominently in studies of social complexity in the early Greek Bronze Age.

II. Methods

Economic organization is an integral facet of social complexity. A model of the utilization of pigs and other mammals in ancient settlements provides a starting point for the reconstruction of ancient economies. This reconstruction can then be used as a proxy indicator to approach issues of social complexity.

This study provides a novel approach to the examination of social complexity in several respects. At the broadest level, it employs zooarchaeology to address social complexity in Early Bronze Age Greece. In so doing, it examines the most important issue for understanding animal utilization in society - that is the choice to manage one species over another. This research specifically concentrates on the utilization of pigs as the avenue by which economic organization and ultimately social complexity is reached. However in so doing, it necessarily addresses the additional major domesticates – sheep, goats and cattle. The management of these species is addressed through the novel application of the principles of behavioral ecology to the analysis of four Early Helladic faunal assemblages – Helike, Lerna, Tsoungiza and Tiryns. The conjunction of the two approaches will allow this study to come full-circle and ask how the choice to manage one species over another informs social complexity.

This study combines traditional zooarchaeology and biological theory to address pig utilization in Bronze Age Greece, and ultimately to the study of spatio-temporal aspects of social complexity. Socio-economic and biological relationships between pigs and humans differ from that of other domesticates - and pigs are virtually unstudied. These data and methods have never been combined in the region, and rarely elsewhere. This goal necessitates the accomplishment of the following objectives:

- comparative analysis of faunal pig data from both large and small settlements using differences in percentages of fauna, skeletal part frequencies, and evidence of utilization of different animal products such as skin, meat and bone;
- comparison across domesticates of life history variables and other behavioral ecological data to understand the unique relationships between pigs and humans;

- analysis of the social and economic role of pigs in small-scale Bronze Age settlements.

III. Structure of the Examination

This structure of this examination resembles the life history of an organism. Part I provides the background information (the "genetic code") necessary to understand the evolution and unfolding of the question. Part II is the difficult adolescent stage. It contains an exposition, and details of, the study of the faunal data from four Early Helladic assemblages in the Peloponnese: Helike, Lerna, Tsoungiza and Tiryns. Part III is the maturation of the topic. It draws on the background information provided in part I and uses it in conjunction with the concrete data from part II to provide the final analysis and hypotheses as to the nature of social complexity in Early Bronze Age Greece – and the vital role of pigs in the reconstruction of socio-economic aspects of social complexity.

Part I begins with a synopsis of the Greek Bronze Age (Chapter 2). It provides an overview of the period, discusses issues of terminology and chronology, and concludes with a more in-depth examination of social organization in the Early Helladic period in particular. Chapter 3 addresses several theoretical perspectives of social complexity – craft/economic specialization, social stratification, political organization, and urbanization. Chapter 4 contains an in-depth look at faunal methodology and the ways in which it can be employed to reconstruct aspects of social complexity. In particular it details the way in which social complexity can be seen zooarchaeologically, examines faunal methodology, and addresses its utility in addressing inter-site socio-economic relationships. Part I concludes with a focused examination of pigs, Chapter 5. In particular, this chapter examines the importance of pigs and pig utilization as an aid to understanding social complexity. It lays out in detail particular hypotheses concerning the utilization of pigs in ancient settlements, and discusses issues of domestication and management.

Part II is devoted to zooarchaeology and faunal analysis. It begins with a history of research and the current state of knowledge of zooarchaeology in Greece (Chapter 6). This chapter provides a summary of the extent to which faunal analyses have factored into archaeological investigations, discusses relevant publications on the topic, and addresses the major fauna of the region. Chapter 7 presents the hard data on which this research rests. It details the faunal data from the Early Helladic assemblages at Helike, Lerna, Tsoungiza, and

Tiryns. The data from Helike are derived from my own work as a faunal analyst for the ongoing excavation at this site, while the other data come from my examination of museum materials from previously excavated sites as well as data from site reports.

Part III draws on the background from Part I and the physical data from Part II to illustrate how pigs act as an indicator of economic organization, and thus as a proxy for social complexity. It opens with chapter 8, an analysis of pigs and their role in social complexity in Early Helladic Greece, in which the model, hypotheses, and conclusion are laid out. Chapter 9 addresses behavioral ecology and the economics of animal use, discussing the role of ecological, environmental, and biological factors in ancient animal economies. It explains the integral relationship between behavioral ecology and management/domestication, and concludes by illustrating the way in which the principles of behavioral ecology are integral to the pig "litmus test" for social complexity. The dissertation concludes with chapter 10, a brief "ode to the pig" and its ultimate role in ancient social complexity.

IV. Summary

Archaeology in Greece has a long tradition – and yet, faunal analyses have only recently begun to play a role in reconstructing the ancient history of the region. In Bronze Age Greece alone, a wealth of comparative studies have been undertaken, including mortuary analysis (Pullen 1994), lithics (Kardulias 1992), pottery (Attas et al. 1987), and architecture (Hägg and Konsola 1986). To date, however, no comparative studies with a focus on faunal analysis have been carried out. This research will illustrate the utility of comparative faunal analyses and their potential for addressing social complexity. It will also elucidate the relationship between pigs and socio-economic complexity in Bronze Age Greece.

The marriage of zooarchaeology and behavioral ecology is a unique application to the study of social complexity in Bronze Age Greece. It is the broad theoretical framework in which this analysis is situated. A traditional zooarchaeological approach may focus on the different products the animals produce, while a behavioral ecology approach may focus on the evolutionary differences between cattle and pigs as an explanation for the striking differences between societies making intensive use of one species versus the other. Does this decision affect, perhaps limit or potentiate, social stratification within a society? Pig exploitation is fundamentally different from that of cattle, sheep and goats in ways that exploit the evolutionary differences between the species. Further, both the advantages and disadvantages of pig utilization are generally different than for other domesticates.

This research indicates that the intensity of pig utilization differs between larger, administrative settlements and smaller, village sites. What factor(s) account for this variability? Is this variability shaped by the economic and/or social needs of the settlement? Or is this variability governed by nature? Do the biological differences between suids and bovids, such as time to sexual maturation, litter size, life expectancy, dietary differences and ecological variables help explain the choice to exploit one with more intensity than another? What about other economic variables, such as the intensity of grain/cereal agriculture? Both modern and ancient evidence illustrates the capacity of pigs to destroy crops. For example, in areas with heavily cultivated land for intensive agriculture, pigs may not be a practical domesticate, as they are difficult to keep out of large areas. In smaller settlements with garden cultivation, pigs can be kept out with greater ease, and in addition, they can be left to forage on their own outside the settlement, being harnessed only when needed. Thus, the frequency of pigs in faunal assemblages may indicate not only settlement size, but also the relative size and nature of the local economy.

Each of these questions will be explored in the body of this work. They will be addressed through the exploration of social and economic structures in relation to domesticates across different sites, through social and economic differences between domesticates from a behavioral ecological perspective; and, as per above, a novel look at pigs in particular. In the end, this research will illustrate how pig utilization in ancient settlements can be used to reconstruct economic organization, and transitively, social complexity in Early Helladic Greece.

CHAPTER 2

THE GREEK BRONZE AGE

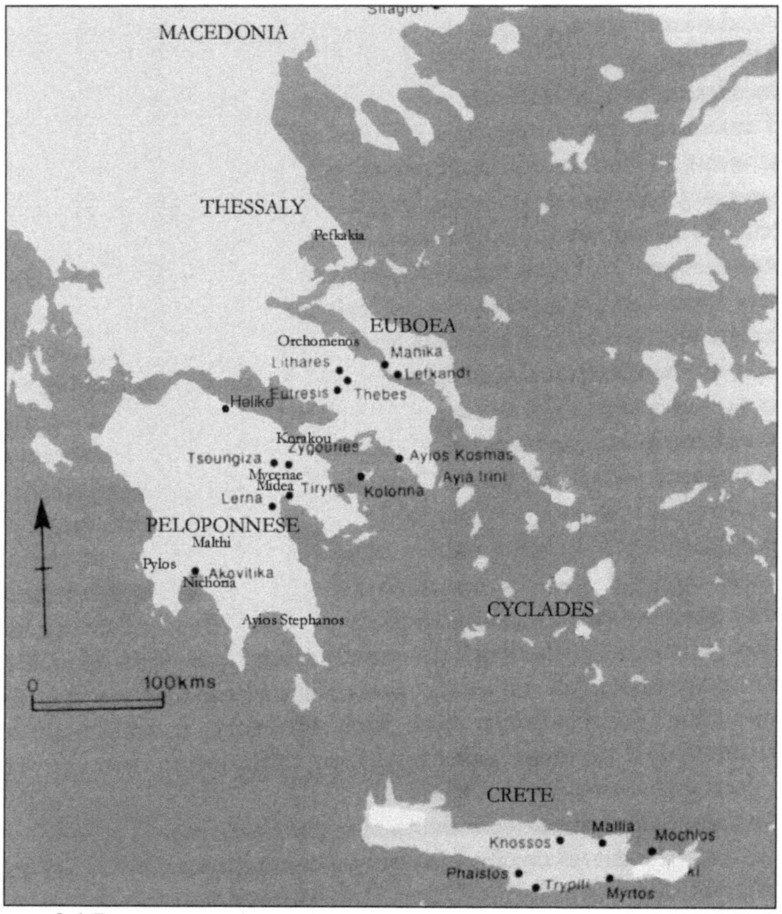

Figure 2.1 **Bronze Age sites mentioned in the text**

Evidence from architecture, the spatial distribution of settlements, and material remains such as pottery and personal affects, paint a varied picture of the Greek Bronze Age (referred to as the Early Helladic period on the mainland). In general, Early Helladic I (EH I) opens with a small start, and Early Helladic II sees the appearance of more complex social systems, as seen in the appearance of corridor houses, clay seals and monumental stone cut tombs with grave goods. This new social system then appears to collapse, with widespread evidence for destruction present at several Early Helladic II sites, ushering in a less-centrally organized Early Helladic III. Whether "invaders", new waves of immigrants, or some other factor, caused the demise of many sites, is difficult to know. However, several questions still remain to be answered about the period which the traditional lines of evidence have not answered definitively. Thus, there is Daniel Pullen's suggestion that the Early Helladic II period was a loosely arranged chiefdom (Pullen 2003), but what of the nature of inter site relationships? Did Lerna and Helike communicate? What type of trade systems were in place? Were some sites subordinate to others? What was the social organization intra settlement like? Was there variability between sites excepting of architecture?

The answers to these questions must be sought via different avenues than those traditionally taken. Though many analyses have been done, faunal analyses are still largely neglected. Of the sites discussed here, Lerna has extensive published faunal data (Gejvall 1969), and new data from Tsoungiza are in preparation (Halstead, in press). Faunal data from Tiryns rounds out the list of published faunal analyses on Early Helladic sites (von den Driesch and Boessneck 1990). The existence of only three published reports highlights the need for more analyses. The remainder of this dissertation will be concerned with clarifying and illuminating aspects of Early Helladic socio-economic complexity via the application of zooarchaeology. It will serve two very important functions:

- It will conduct the first application of faunal analysis to socio-economic complexity in Greece and,
- It will illustrate the utility of such endeavors to future scholars, in the hope that faunal analysis will no longer be viewed as an elective pursuit.

Before proceeding to the fauna, however, a general discussion of the Greek Bronze Age must ensue, addressing important issues of chronology and major sites, social organization, and a brief summary of the period. This discussion will focus on the Early Bronze Age on the mainland in the Peloponnese, but in so doing will discuss where necessary, generalities covering the rest of Greece.

I. Overview

The Early Bronze Age in Greece is an enigmatic period; this is in part because its study has often been overlooked in favor of the periods both preceding and following it. The Neolithic has garnered copious amounts of attention, as has the Late Bronze Age (Mycenaean Period), and to a lesser extent, the end of the Middle Bronze Age. The paucity of research in the Early Bronze Age can be explained by excavation of, and knowledge of, few sites, and a resulting impression that the period holds little in the way of social organization and centralized society – topics of considerable interest to researchers. Study of the preceding Greek Neolithic has garnered a dearth of research primarily because its study falls into the field of prehistorians, commonly the domain of anthropologists, whose interest tends to focus more intensely on smaller-scale societies.

The question that follows then, is why has the Early Bronze Age not been studied as the Neolithic has, since it too falls into the category of prehistory? Part of the answer to this question forms the backbone on the mystery that is the Early Helladic Period – that is, Early Helladic horizons are frequently missing. Excavations of settlements in Greece very often possess a stratigraphic layer cake of continuous occupation. One frequently encounters a rather neat sequence of Neolithic, Mycenaean, Archaic, Classical, Roman and Byzantine horizons literally piled on top of one another. A "missing" Early Helladic horizon may be due to the robbing of material by later societies, or the possibility that Early Helladic sites do not fit into this continuum of occupation. Is it possible that settlements in the Early Bronze Age were located elsewhere, and if this is the case, why?

This chapter will discuss Early Helladic culture on the mainland in order to address these questions and lay out a framework for understanding social organization is this "dark" period. This dissertation will also show that conceptions of a "dark" Early Helladic Greece may be erroneous. The Neolithic period is commonly accepted as a time defined by small, egalitarian settlements, lacking complex hierarchical social organization, whose scattered settlements may have encouraged the development of regional cultures (Dickinson 1994). While the Late Bronze Age, or Mycenaean Period (on the mainland), is often thought to be the first complex civilization in Greece. How did ancient Greek peoples pass from a lack of central organization to a highly stratified, centrally organized, powerful Mycenaean Kingdom? The intervening Early Bronze Age has the potential to address this question, asking an additional question: What if complex social organization actually began in the Early Helladic Peloponnese?

II. Terminology and Chronology

Early Bronze Age scholarship is a web of conflicting chronologies based on type sites in differing regions. It is frequently difficult to assign absolute dates to phases within these periods and even more difficult to compare chronological phases between sites. While there is a need to clarify and standardize some of the terminology used in discussing these periods in Ancient Greece, as well as to briefly address the issue of chronological organization, this study will avoid delving into this tangled web of contentious dates. It will instead summarize some of the major lines of thought on the matter, ultimately creating a working, simplified chronological scheme that serves as a suitable framework for use in this research.

The term Bronze Age is a general temporal category that refers to the time span falling between the Neolithic and Geometric periods, and it extends from 2800 B.C. to 1000 B.C.[1] (Table 2.1).

Period	Date
Neolithic	6800-2800 B.C.
Bronze Age	2800-1000 B.C.
Proto-Geometric (Sub-Mycenaean)	1000-800 B.C.
Geometric	800-600 B.C.
Archaic	600-400 B.C.
Classical	400 B.C.
Hellenistic	300-100 B.C.
Roman	100 B.C. on...

Table 2.1 **Simplified chronology of Greece**

The Bronze Age is subdivided into three broad temporal divisions: Early, Middle, and Late. Within each of these subdivisions, additional temporal categories are employed. These are often the subject of variation and debate (Dickinson 1994; Renfrew 1979; Warren and Hankey 1989). This study avoids the complicated web of temporal and chronological debate, and adopts the most basic of temporal divisions. Thus, the Early and Middle Bronze Ages can be further separated into Early Bronze I, II, and III. The Late Bronze Age is often subdivided into I, II, IIIA, IIIB, and IIIC.

The term "Bronze Age" is a general temporal designation which refers to disparate geographic locations across Greece, including the Greek Mainland, the Cycladic Islands, and Crete. Thus, there is a basic Minoan-Cycladic-Helladic division. The term "Helladic" refers to the Bronze Age solely on the Greek Mainland. Therefore, when discussing the Early Bronze Age in the Peloponnese on the mainland, the term "Early Helladic", or EH, is employed. The Late Helladic is alternatively known as the Mycenaean Period on the mainland. On Crete, the term Minoan is employed, and so the Middle Bronze Age becomes the Middle Minoan (MM), and the Late, the Late Minoan (LM), etc. The Cyclades too have distinctive terminology, utilizing the term "Cycladic" for their temporal divisions. Thus EC, MC and LC are employed for Early Cycladic (Early Bronze), Middle Cycladic (Middle Bronze) and Late Cycladic (Late Bronze) (Table 2.2).

---

[1] Debate exists concerning the opening date for the Greek Bronze Age. The date of 2800 B.C. given is arguably the most commonly accepted, however others have proposed an earlier date of 3300 B.C. For further comment on this debate, see Dickinson (1994:18-19).

Bronze Age temporal divisions have been the subject of extensive debate in large part because they stem from seriating pottery.[2] Chronological sequences for Mainland Greece were based largely on ceramic seriation from various sites (Figure 2.1). Early Helladic I chronology is based primarily on pottery typologies from Eutresis and Lithares (Boeotia), Lake Vouliagmeni (near Perachora), and Palaia Kokkinia (Attica). Early Helladic II chronology is based on Lerna (Argolid), Lefkas, Agios Kosmas (Attica), Orchomenos, Akovitika, Eutresis, Zygouries, Nemea, and Pefkakia.

Date B.C.	Crete (Minoan)	Cyclades (Cycladic)	Mainland (Helladic)
3300			
3200			
3100	**EM I**		
3000			
2900			
2800	**EM IIA**		**EH I**
2700			
2600		**EC**	
2500			**EH II**
2400	**EM IIB**		
2300	**EM III/MM IA**		**EH III**
2200			
2100			
2000			
1900	**MM IB/IIA**		**MH I**
1800	**MM IIB/IIIA**	**MC**	**MH II**
1700	**MM IIIB**		**MH III**
1600	**LM IA/IB**	**LC I/II**	
1500			**LH I, IIA**
1400	**LM II, IIIA1**	**LC III**	**LH IIB, IIIA1**
1300	**LM IIIA2, LM IIIB**		**LH IIIA2/IIIB1/B2**
1200	**LM IIIC**		
1100			**LH IIIC**
1000	**Subminoan**		**Submycenaean**

Table 2.2 **Chronological correlations for Bronze Age phases throughout Greece**

Early Helladic III chronology is based fundamentally on Eutresis, Thebes, Lerna, Aigina, Raphina, Tiryns, Lefkandi and variously Orchomenos.[3] Dating of the Middle and Late Bronze Age follows a similar methodology, with this reliance on pottery for dating

[2] The tripartite dating system for the Aegean was first proposed by Arthur Evans during his excavation of the Palace of Knossos on Crete from 1921-35 (see *The Palace of Minos at Knossos I*, 1921). It was later applied to the mainland by Wace and Blegen in 1918.

[3] Warren and Hankey (1989) offer a detailed discussion of each of the sites contributing to the Early Bronze Age chronology in all areas of Greece, including the Cyclades and Crete.

sequences extending to later Mycenaean ware. Note of the sites contributing to Early Helladic chronological sequences is made in order to address the major settlement areas. As this research has as its focus the Early Bronze Age, discussion of sites affecting later chronologies will not be made. Warren and Hankey (1989) are an excellent resource for this information.

Whilst sequencing pottery is certainly a way in which to distinguish between different temporal traditions, and arguably cultures, there are numerous problems with relying solely on this method. Foremost among these is the need to force all subsequent sites into an artificial chronological framework. This raises the issue of absolute chronologies based upon scientific methods, such as radiocarbon dating, and the merit of such methods. It also highlights the need for alternate methodologies, among which faunal analysis holds great potential.

Bronze Age Greek chronologies have been shown to be problematic; however resolution of this problem is beyond the scope of this work at hand. In the name of clarity, this analysis adopts the most basic divisions within Bronze Age Greece, thus utilizing the basic temporal categories Early, Middle and Late. Further, it has as its focus the Early Bronze Age on the mainland, or the Early Helladic (EH) Period, and will use the general temporal span from 2800-2000 B.C. Where necessary, subdivisions within the Early Helladic period will be utilized (I, II, III). However, this analysis will abandon absolute dates associated with these subdivisions, as they cannot be clearly shown to be the same at all sites of Early Helladic date. Further, absolute dating is not of vital importance to the central aim of this work – the reconstruction of social complexity via faunal analysis in the Early Helladic Peloponnese.

III. Social Organization in the Bronze Age: A Summary

The following will provide a general synopsis of social organization in the Greek Bronze Age. It is not meant to be an in-depth analysis of all major periods and phases, but rather is meant to serve as a framework within which to discuss in greater depth and detail, the Early Helladic period.

Early Helladic:

If the Neolithic is generally recognized as the period ushering in animal domestication and the beginnings of early settlements in Greece, the Early Helladic I period (ca. 2800-2500

B.C.) opens by building on this model. It is a period characterized by small villages typically on low ground near coastal areas. The change from the Neolithic to the Bronze Age on the mainland has been attributed to new peoples arriving from Asia Minor. While this influx has been variously termed an invasion, simple migration can not be ruled out (Forsen 1992). It is thought that these new people were the pre-Hellenic population of Greece. The Early Helladic I period sees the appearance of new villages and evidence of trade with the Aegean islands, especially Crete. The southern mainland yields evidence of a more advanced material culture than do Northern Greece and Thessaly.

Toward the middle of the early Helladic, EH II (2500-2300 B.C.), evidence exists for the beginnings of monumental building in the form of possible administrative complexes at Akovitika and Lerna (House of the Tiles).[4] Rising prosperity is also suggested by larger settlements at Zygouries and Tiryns, at which gold and silver jewelry was found in tombs. However it is during the later Early Helladic, EH III (2200-2000 B.C.), that the period yields several interesting aspects. At nearly all Early Helladic II-III transition sites, evidence of destruction is present. This destruction generally takes the form of burning, and in several cases appears to coincide with site abandonment (as may have been the case at Helike and Lerna). Some have attributed this destruction to hordes of invaders, although this is a theory that is gradually losing credibility.[5] The destruction phases coincide with the introduction of a new material culture, characterized by Minyan Ware (also termed Orchomenos ware), which is a fine, wheel-made pottery. Settlements of Early Helladic III date contain significant quantities of this wheel-made ware, ushering in the beginnings of a new technology which would eventually lead to mass production and craft specialization. The beginning of trade specialization is apparent with the possibility of specialized centers of pottery production. It is debated whether this change in material culture represents an invasion of Greek speakers into the region. The intrusion of the Greeks into the region has been variously dated from 2200 B.C. to 1500 B.C.

Middle Helladic:

The Middle Helladic period (2000-1500 B.C.) is witness to an initial continuity of patterns of culture established in the preceding period, until the eventual emergence of the Mycenaeans and Minoans. Settlement patterns exhibit differences from the Early Helladic period, however, in that they are typically located on top of rocky hills or eminences. In

---

[4] See Hägg and Konsola (1986) for a discussion of architecture at both Akovitika and Lerna.

[5] J. Forsen has shown that there is no consistent pattern to the destruction of EH II sites, with many, such as Korakou (near Corinth) displaying EH II-III continuity (Forsen 1992).

contrast to the dispersed nature of Early Helladic settlements, Middle Helladic settlements appear to be nucleated with a relatively low overall site density (Dickinson 1994), with sites mainly concentrated throughout the Peloponnese and central Greece. Although few sites survive intact, Malthi in the southern Peloponnese (Messenia) stands, at this point in time, as the one most completely excavated, with the eventual publication of Lerna V anticipated to become the "type" site. The settlement at Malthi appears to have been a fortified village, lacking in any central social organizational plan.[6] Malthi may function as a template for at least some aspects of Middle Helladic society; however there is some contention concerning its Middle Helladic date, with some suggesting that since fortified settlements are not characteristic of the beginning of the Middle Helladic, it should instead by give an Early Helladic III date (Dickinson 1994). In general, many of the best known Middle Helladic sites have been shown to have continuity from the Early Helladic, including Lefkandi, Lerna, Eutresis and Aigina Kolonna.[7] Additional sites rounding out Middle Helladic assemblages include Nichoria, in the southern Peloponnese, and Ayios Stephanos.

From the limited data available, the end of the period is marked by a rapid rise in wealth and an increasing complexity in social organization associated with a palace-based civilization that sees its true flowering with the Late Helladic Mycenaean kingdom. The presence of fortification walls at several Middle Helladic sites suggests the beginnings of increased belligerence between locales and may form the precursor to the heavily fortified Mycenaean state. In Messenia fortification walls have been unearthed at Malthi, Peristeria and Pylos, as well as at a handful of additional sites in Attica. The Middle Helladic appears to have been influenced by the Minoan civilization on Crete, and is witness to the first shaft graves at Mycenae (Grave Circle B). The rise of Mycenaean culture begins during the late Middle Helladic, centered initially on the eastern Peloponnese and central Greece.

Late Helladic:

It is the Late Helladic or Mycenaean Period (1500-1000 B.C.) that has garnered the most attention, with the spectacular citadels of Mycenae and Tiryns among the best known. Late Helladic culture may be viewed as the harbinger of the later power that becomes Greece, and is certainly the most complex period of the Aegean Bronze Age. It is a palace-based

[6] See M.N. Valmin (1938) *The Swedish Messenian Expedition: Malthi Dorion*

[7] Though the majority of the pottery from Kolonna is MH, the settlement is sufficiently different in terms of size and wealth from any MH mainland sites so that there is an argument for considering Kolonna its own category. For an in-depth discussion see Zerner et al. (eds.) 1993.

society with complex social organization, stratification and hierarchy. Grave goods recovered from Tholos tombs suggest royal wealth, prosperity, and numerous small kingdoms. It sees craft specialization with palatial administrative centers controlling much of the wealth and resources of the countryside, coupled with an increase in population density on the landscape. As with the rise of most complex societies, the Late Helladic period is also marked by political unification and military aggression. Homer's Trojan War is thought to have occurred during the Late Helladic III period. The Mycenaean period also contains one of the earliest forms of writing in Greece, Linear B script, found at Pylos on the mainland and Knossos on Crete. Palatial records were kept in Linear B detailing trade goods and palatial holdings. They are a rich source of knowledge for Late Bronze Age economic systems, and much research has been done with them, beginning with their initial decipherment and study by Ventris in 1953.[8] In the realm of zooarchaeology, the Mycenaean period is also well documented, with a recent focus on feasting (Bendall 2004; Halstead and Isaakidou 2004; Wright 2004). The archaeological complex at Mycenae in the Argolid is perhaps the most well known Late Helladic site, and as been used extensively in social reconstructions of the period. Though additional sites are scattered throughout the Peloponnese, the center of the Mycenaean kingdom was in the Argolid, near modern day Argos, and is populated by several fortified settlements/centers, among which are Tiryns, Midea and Mycenae itself. Much of the additional information about Late Helladic Greece comes from tombs and cemeteries, such as Mycenae Grave Circle A, Lerna's shaft graves' fill and Prosymna tombs 25 and 26; settlements at Korakou, Messenia, and Ayios Stephanos round out the Late Helladic grouping (Warren and Hankey 1989). Massive amounts of material have been published on the subject of the Late Bronze Age in Greece, and it is for this reason that this summary is necessarily brief.

IV. Early Helladic Social Organization in Detail

As this research is concerned with social organization in Early Helladic Greece, a more nuanced discussion of the period is needed. While an in-depth analysis of social complexity is addressed in Chapters 3 and 4, a summary of what is commonly accepted on the basis of traditional lines of inquiry is here set out. The following pages will examine

[8] First discovered by Evans at Knossos, Linear B script was later deciphered by Michael Ventris. For detailed discussion of Linear B see Ventris and Chadwick (1953) and Chadwick (1967).

various categories of archaeological evidence found in early Helladic contexts in order to set out a framework for what is known about the period. This framework will later be compared to that suggested by the faunal data (Chapter 8).

Social organization can loosely be described as a way of differentiating among individuals or groups within a society. Three broad categories of evidence can be employed in its reconstruction, and these are categories that address the individual, the group and the population. These three categories can be subdivided into several sub-categories to facilitate greater clarity and understanding of a period. Thus, social differences between populations can be viewed by examining the spatial distribution of sites across a region, as well as variation in site size, craft/economic specialization and architectural differences. Social differences between groups can be examined by looking at intra- and inter-site differentiation in order to see if variability exists in built structures. Finally, social differences among individuals may be examined where possible through analysis of small finds such as stone, metal, and bone (items frequently accompanying burials and associated with graves). As many of these ideas overlap and are intertwined, the remainder of this chapter will move fluidly between these broad categories in order to paint a picture of Early Helladic social organization.

The Population/the group/the individual:

An examination of the variability between settlements in all phases of the Early Helladic period will address central aspects of social organization by examining these differences with an eye toward addressing variability among Early Helladic populations, groups and individuals. The accomplishment of this task will rest upon delineating differences in the spatial distribution of sites, variability in settlement size, architecture and craft/economic specialization.

Spatial distribution: In general, the opening of the Early Helladic period (EH I) is witness to an increase in settlement size as compared with the Final Neolithic. Early Helladic I sites appear to be concentrated in coastal areas, such as southern Attica and the nearby islands of Euboa and Aigina. They are settlements that are generally situated near the coast in lowland areas, often with available land for small-scale agriculture. Little is known about this early period, perhaps due to excavation bias such as the inability to excavate deep enough, but also perhaps due to the tendency of later Early Helladic II peoples to demolish earlier settlement phases in the construction of new structures. Tsoungiza in Ancient Nemea is one of the few projects to have reached these early levels (Pullen 1986, 1988, 2006 in press). Tsoungiza illustrates a probable continuity between Early Helladic I and Early Helladic II phases of occupation.

Sites of Early Helladic II date are more numerous, and yield further information about the spatial distribution of Early Helladic settlements. The Early Helladic II period is often typified by the Lefkandi I culture (EH IIB) whose major sites include Pefkakia (coastal Thessaly), Lefkandi and Manika (Euboa), Raphina (Attica), Eutresis, Orchomenos and Thebes (Boeotia), and Kolonna (Aigina). Evidence from this period suggests increasingly larger settlements that may have been administrative or regional centers. Lerna in the Argolid is the best example of such a site. The House of Tiles at Lerna dates to this period, and provides one of the best preserved examples of a Corridor House (several rooms and a second storey). The House of Tiles has been variously interpreted as a regional administrative center and/or palace. Interpretation aside, its existence illustrates a new variability both inter- and intra-site with respect to social differentiation. This is an area to which further attention will be directed shortly.

Settlement size: Daniel Pullen has argued for the beginnings of a hierarchy of settlement size beginning in the Early Helladic (2003). By Early Helladic II he sees three types of settlements, based upon size: small, medium, and the larger "super" site or administrative center. By analogy with other societies and cultures, a general correlation can be made between the size of a site and its relative social and economic importance. Thus, Pullen sees a hierarchy of settlement size as suggestive of social and economic differences in a society – and perhaps the domination of smaller sites by the larger ones. While inter-site size differentiation may well be indicative of the beginnings of complex social organization, the relationships between settlements of similar date is not well understood. Ceramic evidence indicates that certainly exchange and possibly, trade, were occurring, however the precise nature of the interrelationships is still opaque. One of the aims of this research is to examine whether faunal analysis can illuminate this area. This is an issue that will be given attention in Chapter 8.

Architecture: An array of architectural forms appear in the Early Helladic period, among which are apsidal houses/structures and longhouses (alternately referred to as megara). These structural types first appear in the Early Helladic II and become standard types in the ensuing Early Helladic III period. Several have argued that these architectural forms illustrate an "Anatolianizing" of Early Helladic culture, lending support to the idea of a wave of migrants from the East at some stage in the period. The most well-known structure comes from the Early Helladic II period, and this is the aforementioned Corridor House, of the type found at Lerna. Corridor houses can be defined as buildings with two or more large rooms on the ground floor, flanked by long narrow corridors, some of which have staircases providing access to a second storey. Their exact functions are still the subject of debate, but it is commonly accepted that they may variously serve as a combination of public gathering and administration, private living and controlled storage (Hägg and Consola 1986). Corridor houses are overwhelmingly accepted as at the very least, indications of central administrative sites. Their existence is also one of the strongest forms of early evidence for the beginnings of larger scale, administrative areas. They are found at several sites throughout southern Greece, and although of similar plan, they vary in scale.

In addition to these large-scale buildings, small rectangular structures, possibly representative of small-scale vernacular settlements, are present. The architectural evidence from Early Helladic settlements appears to support Pullen's idea of a tiered settlement hierarchy within which are found settlements of varying size and importance. Architecturally, the Early Helladic period in the Peloponnese appears to have at the very least, the beginnings of a socially complex society. This is further supported by evidence from craft specialization and trade.

Craft specialization and trade:

> Craft specialization involves not only economic differentiation (i.e., division of labor), but also economic individuation, in which certain persons spend a significant portion of their time in particular non-subsistence activities that do not engage others. As Evans has noted, specialized production requires that 'these specialists cannot participate in some or all of the basic community subsistence goods through exchange for their craft products' (Kardulias 1992:439).

When addressing questions surrounding variation in settlement size as an indicator of social complexity, one of the qualitative measures that may be used is the presence of finds that are indicative of craft specialization and/or trade. These include seals or sealings, pottery, obsidian, metal and agricultural products. If such materials are found within settlements, their presence may be taken as an indicator of the existence of trade networks and communication. A solid body of evidence from Early Helladic settlements supports craft specialization, and trade, during this period. The following pages will address this corpus, beginning with the evidence from sealings, and covering additional areas of agriculture, pottery production, obsidian and metal. A theoretical discussion of craft specialization as a part of social complexity is addressed in greater detail in Chapter 3.

Sealings: Sealings, though relatively rare, are a significant enough occurrence in Early Helladic settlements to command a study of their own. Their existence suggests a network of craft specialists who were very likely engaged in trade. At The House of Tiles at Lerna excavation uncovered over 100 of these clay seal impressions. The exact function of the seals is still often debated, but they are commonly interpreted to indicate ownership, possibly functioning to seal amphora, vessels or containers carrying textiles and other commodities. In addition to Lerna, a large number of Early Helladic II settlement contexts (and two burial sites) in the Peloponnese and southern Greece have evidence for seals, sealings and seal impressed objects (17 have seals, 6 have sealings, and 11 have seal impressions) (Pullen 2003). At the site of Geraki in Laconia (southern Peloponnese), there is evidence for nearly 50 sealings, discovered in a room destroyed by fire, much like Lerna (Weingarten et al. 1999). The backs of several seal fragments from Geraki contain textile imprints. Analysis of the imprints suggests the cloth may have been linen or at least finely spun material.

> ....though the evidence is (inevitably) indirect, it is entirely possible that textiles of relatively fine quality were being produced on the mainland of Greece in Early Helladic IIB. Indeed, it now seems likely that a fully-formed linen industry had already come into being – a thousand years before the one so richly documented at Mycenaean Pylos (Weingarten et al. 1999:364-5).

In addition to their obvious role in trade, seal impressions also provide evidence of the spread of cultural ideas, and thus contact between different regions. Geraki's sealings hint at culture contact, as some of the sealing motifs have Anatolian antecedents such as the swastika. Other sealings are similar to Early Bronze Age cylinder seal impressions from Beit Yenah, Palestine (Weingarten et al. 1999). These impressions may provide some of the best evidence for a trade network that extended from the Peloponnese to Anatolia. Perhaps Lerna was a stop on a trade route with Anatolia, and perhaps Geraki was also on this route. Evidence from the southern Peloponnese suggests a north-south trade route for metal and obsidian. It is feasible that Geraki could have been a production center for textiles (linen), and textiles normally ride on the back of metal trade. The textiles could then have been sent north to Lerna, who then sent them on to Anatolia (Weingarten et al. 1999:370). Economic complexity and craft specialization, coupled with long-distance trade and social contact is certainly indicative of a socially complex society.

Raw Materials: The evidence for trade and craft specialization extends beyond the use and existence of sealings, however. Manika on Euboa, variously dated to Early Helladic II and Early Helladic III has yielded evidence for its role as a possible production and trading center (specializing in obsidian, metals and grains) (Sampson 1985). Interestingly, there is no evidence for monumental architecture such as megara or corridor houses, and this factor has been interpreted to mean that the site lacks evidence for public buildings. Manika does, however, contain a large cemetery with burials interred in monumental rock-cut tombs; this elaborate burial treatment is unparalleled in Early Helladic II, and may provide evidence for a well-organized social hierarchy and social differentiation between individuals (Sampson 1987). The fairly elaborate evidence from sites like Lerna, Geraki and Manika is all the more interesting, when viewed against the general paucity of additional known sites in the Peloponnese of significant size.

Analysis of trade in raw materials during the Early Helladic period further supports the theory of the existence of a fairly well-developed socio-economic network between settlements. Evidence of the importation of lithics, like the fine, parallel-sided flint blades in the Early Bronze Age, coupled with the existence of obsidian from the island of Melos is "evidence of trade or at the very least exchange" (Kardulias 1992:433). Kardulias (1992) has drawn the following conclusions regarding Early Helladic social and economic organization from data derived from lithic analysis. He suggests the following:

> 1. The procurement of raw materials in southern Greece was embedded in other activities, probably the acquisition of food and truly precious commodities such as metals;
> 2. The quantity and distribution of material at Agios Stephanos and the Southern Argolid suggests the presence of full-time specialized stone knappers in the Argolid, but only part-time artisans at Lakonia (Kardulias 1992:433).

Thus lithic analysis appears to yield similar conclusions regarding the existence of trade networks and socio-cultural contact as those suggested by seals and architecture. But what does the most common, and basic, of all economic areas – agriculture – suggest?

Agriculture: As of yet, sufficient botanical data for the Late Neolithic through the Middle Bronze Age has not been analyzed and reported in great enough detail to enable the production of detailed agricultural models for the Early Helladic period (Hansen 1988). However, enough data exists to support the conclusion that there was a slight increase in crop diversity with the addition of the vine in Early Helladic Greece. Renfrew (1972) proposed a model for the development of a redistribution system of subsistence commodities that "emerged as a consequence of intensive exploitation of a new spectrum of food plants, notable tree crops, yielding a new diversity in produce" (Renfrew 1972:480). He suggested that wheat, olive, and vine were the main cultivars in the Early Helladic Mediterranean, along with an increased number of cultivated legume species. This diversification of crops may have allowed more marginal land to be exploited, resulting in increased production and greater security for farmers. Specialization in certain crops could then have arisen, accompanied by the production of a surplus that could be exchanged for other products. This would allow for redistribution system to arise as a mechanism for moving surplus of one product into areas lacking that product (Hansen 1988).

Halstead (1981) articulated a concept of social storage which is related to this idea of the existence of a redistributive system. He posited that social storage would function as an exchange of surplus in time of need with the expectation of future reciprocity. This type of exchange would buffer the communities against periodic shortages brought on by crop fluctuations. Diversification of species or habitats exploited would also be necessary to ensure a surplus production of exchangeable goods. Redistributive centers could have arisen in this context, ultimately controlling the exchange of surplus from one region to the next (Hansen 1988:41).

Halstead's social storage idea, coupled with Renfrew's redistribution theory, provide areas that hold potential for future study of Early Helladic social organization. While sufficient botanical evidence is still lacking, Hansen (1988) saw an increase in legume species from Near East in the Late Neolithic / Early Bronze Age, such as chick peas and horse beans. The existence of these species can certainly be viewed as further evidence of contact, exchange, and probably influence from societies outside the direct geographical sphere of Early Helladic Greece. The introduction of the vine, as seen at Lerna and other Early Helladic settlement contexts, further supports the theory that during the Early Helladic period, the Peloponnese was in the throws of developing social complexity.

Ceramics: Discussion has already ensued of the pottery sequence that set forth the "accepted' Early Helladic chronology, and thus will not be reiterated here. A brief sketch of some general trends in ceramic assemblages is useful, however, as an aid that furthers understanding of areas of possible trade, culture contact and change. On a very general level, Early Helladic wares are characterized by Anatolian forms, such as the two-handled depas cups, one handled tankards, and spherical pyxides. Prototypes of these wares are found in Troy II/III contexts, and it is probable that the potter's wheel also comes from the East.

Lefkandi ware which typifies the Early Helladic II-III period falls under this Anatolian influence. While the specific chronology and pottery sequence from each site is important, what is more important is what it suggests. Pottery from the Early Helladic period indicates that goods passed hands from disparate geographic regions. For example, there is a preponderance of Cycladic-style decoration on ware from the Northern Peloponnese, and ceramics from the island of Euboa made their way to the mainland.

V. What the Evidence Suggests....

Archaeological evidence from Early Helladic Greece suggests that complex social organization began during this period. The spatial distribution of settlements shows the existence of larger sites surrounded by smaller, "satellites"; differentiation exists among sites with respect to size, architecture and craft/economic specialization. Each of these forms of evidence paints a picture of at the very least, the beginnings of a socially complex society of the type that eventually was to become the hallmark of the Later Mycenaean State. In order to understand social complexity, however, it is necessary to examine the theoretical components that contribute to the term. How is social complexity defined? What traits characterize a socially complex society? The ensuing chapter addresses these questions, providing the theoretical framework for the eventual application of faunal analysis to social complexity.

CHAPTER 3

THEORIES OF SOCIAL COMPLEXITY

What is social complexity? What are the main factors that constitute a socially complex society? The more intensely this question is scrutinized, the more contentious the definition. Social complexity has been defined and delineated in various ways depending upon the theoretical persuasion of the writer. Anthropologists, historians, sociologists and archaeologists all have a variety of definitions. But is a working understanding of social complexity truly so complex? Theoretical discussions of culture and society are open to intense scrutiny and interminable debate, but the requirements of an "archaeological definition" of social complexity are different than for other disciplines, because the definition has to be couched in terms that relate to the archaeological record itself – in terms of material culture. This analysis, therefore, will break down the term into its various inherent *archaeologically* visible components, and delineate just *some* of the facets which characterize socially complex societies. For the present purposes, one need only develop a minimal working definition of social complexity in the context of Early Bronze Age Greece. This chapter will explore selected theoretical components of social complexity in order to provide a framework in which to place the faunal evidence under analysis.

One of the many ways archaeological correlates of social complexity in a society can be approached is through its simplification into three broad facets:

- Evidence for craft and/or economic specialization
- Evidence for social stratification/political organization
- Evidence of urbanization

Much debate has ensued over the degree to which each of these facets must be present in the archaeological record to indicate social complexity. This study is not a venue for review of these debates, or a platform to provide a critical analysis of each facet. Rather, this chapter provides an understanding of the ways in which social complexity has been conceived, and identified in, ancient societies archaeologically.

I. Craft/Economic Specialization

V. Gordon Childe was one of the first to study the impact of craft specialization in ancient societies. He hypothesized (1958) that craft specialization led to social evolution.

Childe (1930) also proposed that specialization of labor was one of the roots of the European Bronze Age's economic development. He saw the need for a regular system of trade or barter to obtain raw materials and finished products as contributing directly to the breakdown of the self-sufficiency that characterized Neolithic communities; lurking behind this was the idea that the appearance of bronze working was indicative of emerging societal complexity in prehistoric Europe. In Childe's scheme, metalworking was the critical example of craft specialization and led directly to the development of the European Bronze Age and social complexity. Childe viewed "ideas brought in from abroad, and not the social interactions governing the transmission" as agents of culture change (Gilman 1996:67).

In essence, Childe saw economic agents as the driving force behind social evolution – and by this term he meant an increase in differentiation between individuals in both a monetary and political sense. While craft specialization is one aspect of the economic forces which are undoubtedly part and parcel of socially complex societies, Childe's argument is tenuous in his assignation of metallurgy as the driving force behind craft specialization and ultimately social complexity. To attribute the rise of social complexity to one single phenomenon is to ignore the myriad of factors that contribute. The evolution of social complexity, in Childe's scheme, is a linear progression from a simple, egalitarian Neolithic to a socially stratified, complex Mycenaean state. This is an overly simplistic view. Certainly technology is socially embedded and this means craft specialization, but one needs to be more critical rather than postulating, as Childe did, a linear approach to social complexity beginning with craft specialization, specifically metalworking.

> The dominance of commodity exchange is associated with the development of the polis, not in a simple causal relationship, as Engels proposed, but as part of a complex interaction of changes in the forces and relations of production (Gilman 1996:68).

While it is possible that metalworking was a driving force in emerging complexity in some settlements and/or ancient societies, it may not have been the case everywhere. To draw such a wide generalization is to ignore variability in different cultures – and to overlook a host of other variables.

Craft specialization as an archaeological and anthropological phenomenon is complex because there is more than a single factor responsible for its rise. The intricacy surrounding its place in social complexity arises from the different factors contributing to such specialization (i.e. access to raw materials, need for specialized knowledge, degree of capital investment, quantity of goods produced and the distribution system tied to those goods, and the investment of time/labor in production of the goods). Delineating just some of the variables that contribute to craft specialization makes it clear that it must be viewed as a form of economic adaptation to any number of different social and economic circumstances.

Further complicating the issue of craft specialization is disagreement over what exactly specialization is, and how it is seen in the archaeological record. Orthodox archaeological terminology makes a distinction between community/site specialization and occupational/producer specialization. There is substantial ethnographic and archaeological evidence that supports the existence of community specialization among a wide range of non-state societies (Chapman 1996:74). Since craft specialization is vital to the rise of economic complexity, and ultimately to social complexity, a definition will aid in clarifying this discussion. "...Craft specialization can be viewed as the restricted production of specific goods or a range of related goods for use beyond the immediate needs of the producer or his close household or relations" (Rosen 1997:84).

Rosen's conception of craft specialization draws to mind the related issue of discerning between different types of specialization – household or community. There can be community craft specialization which is generally organized at the household level with little suprahousehold control over production. Such community specialization often entails village interdependence in a regional exchange system – so the material correlates of craft specialization may represent a host of organizational structures (Chapman 1996:74). Rather than becoming more of a tangled web, these different types of specialization illustrate the inseparable relationship between craft specialization, economic complexity, and organizational structures (which may be political, administrative or social).

Before exploring the range of additional variables that contribute to social complexity, it is important to clarify that at a certain level craft/economic specialization and its relationship to social complexity has the danger of becoming a "chicken and egg argument".

> Initially it is important to avoid the confusion between specialization as an *indicator* and as a *cause* of increasing social complexity. However one conceives of complexity, functionalists, Marxists, and others would agree that the emergence of complexity subsumes the emergence of specialization, as the organization of labor changes and more inter-dependent productive systems appear. The critical questions concern the degrees and types of such specialization, and the relationship(s) between their appearance and control of wider productive systems, the appearance of hereditary inequality, the expansion of regional political systems, and so on (Chapman 1996:74).

Craft (and economic) specialization did play a central role in traditional formulations of the rise of the state, and thus social complexity. However, it was but one of several factors. Social stratification and political organization are also integral to social complexity, and arguably played an equal part in contributing to its rise. Thus it is to the contentious issue of class and politics that this discussion now turns.

II. Political Organization and Social Stratification

Political organization:

Societies defined as socially complex are often characterized by a certain measure of centralized political organization. Political organization in this respect loosely denotes a system in which power (in an administrative or social context) is in the hands of a few. "Centralized" political organization may mean little more than a society loosely organized into chiefdoms, and it may also denote the rigidly controlled, bellicose society of the Mycenaean state. The degree to which the social group under study is organized politically (i.e. Chiefdom, State or otherwise) is less critical in this discussion than the fact that *some* type of political unit controls *some* aspects of the group. This political organization need not necessarily be centralized geographically, but may be centralized with respect to any one minority subset of a society. For example, a small settlement with a village "head" who controls some aspects of the flow of commodities may be a form of centralized political organization.

> Historians, anthropologists, and archaeologists, have seen the rise of political hierarchy and centralization in a mutually reinforcing linkage with ever increasing specialization in production, especially of elite goods, or goods to which access was restricted for one reason or another. Control of specialized production has been construed as a prerequisite to greater socio-economic control, as well as a means of establishing the symbolic legitimacy of these control systems…That is, the elite products of specialized production confer status. Copper and other metal goods have been especially the focus of such models (Rosen 1997:82).

This simplistic explanation does not imply that a centralized political system necessarily controls the production or distribution of commodities or craft specialization – only that it *may be* a characteristic of such a system. One of the problems with traditional discussions of craft specialization has been the all too frequent assumption that specialized production must be associated with elite control of the economy. There is not necessarily a direct, linear developmental relationship between social stratification, and by extension, political hierarchy, and economic organization. While economic complexity will ultimately become a part of any state level society, it need not accompany political changes immediately (Rosen 1997:83). Endless discussion over which factor leads to the other does not, for the purposes of this analysis, further understanding of social complexity. Rather, what must be highlighted is the fact that some type of centralized, political administration (of which elites and social hierarchies are a part) is an element of a socially complex society. However, the two systems (economic specialization and centralized political control) are clearly related, and often tend to develop concomitantly and ultimately reinforce one another.

Social stratification:

It is difficult to discuss political organization without broaching issues of social stratification, for there is a distinct, and frequent (though not always necessary), relationship between the two. In most cases where one factor is present, so is the other. Thus, in a centralized political system, there are often elites, or those who hold power over others. These elites, or officials, are distinguished from the general inhabitants of a settlement. Eventually, greater social differentiation can occur, and this may lead to increased social stratification within a society. While it is true that egalitarian societies exist, it is more often the case that a socially stratified society is inherent in a strong political center.

Discussing social stratification can be a sticky endeavor. Primate groups have social stratification – but does this make them part of complex societies? Part of the problem lies in how the topic is conceptualized. Social stratification in the context of this discussion must be viewed as one of the interwoven factors that contribute to a complex society. Thus, while primates, and many other species, might live in stratified groups, these groups do not *also* have the additional components that mark a society as complex. Social complexity is the end result of a variety of intersecting, but not necessarily linear, variables that each contributes bits and pieces to an intricate tapestry. This is not to suggest that a complex society will necessarily have each of these variables developed to the highest degree, but it will have aspects of all these variables – some may simply be in their infancy.

Since the Greek Bronze Age has been hypothesized to be characterized by chiefdoms (Pullen 1992, 2003), it is necessary to briefly address this form of political organization, and illustrate how it can be a part of a socially complex society. Chiefdoms are conceptualized of as existing just prior to state level societies. While both chiefdoms and states have highly centralized administrative structures, states tend to be more hierarchical (Cohen 1978). In addition, there is a difference in the autonomy of settlements in the region surrounding this center. In a state society, the regional settlements tend to have less autonomy, while in a chiefdom, local settlements still remain fairly autonomous. Both political systems, however, are characterized by these central administrative centers that do exercise some a measure of political and economic influence. The distinction or delineation of degrees of political influence and control, while important in some analyses, is not central to this discussion. While centralized political systems are certainly a component of social complexity, in and of themselves, they are not necessary to for it to occur. This idea will become clearer with a discussion of urbanization.

III. Urbanization

The study of the rise of cities is a popular subject in all disciplines. Max Weber (1958) brought the study of urbanization to the academic forefront, and his theories have been widely applied to diverse geographic areas such as the modern United States, Medieval Europe and the Classical World. Weber conceptualized urban environments in a variety of ways, but it is his economic definition that is most pertinent to this discussion. "Economically defined, the city is a settlement the inhabitants of which live primarily off trade and commerce rather than agriculture" (Weber 1958:66). Weber continued to write that economic versatility was important and was established by a place of regular exchange of goods (market). In this meaning, the city is a marketplace. He saw the combined forces of a market economy and a highly centralized society as both contributing to the rise of urban areas. Weber did not neglect political systems in this discussion, however. He theorized that cities were also usually the place of lordly or princely residence (in other words, a political/administrative center). Weber identified two main types of cities from an economic perspective - producer and consumer cities; this concept has been used extensively in zooarchaeology and the study of provisioning in ancient societies (c.f. Zeder in particular).

Urbanization plays a key role in social complexity as a concept that unites political, social and economic lines of thought.

> The process of urbanization should then be explained as the sequence in which people develop various institutions for the performance of different functions and services. A growing number of functions and services in a region is often summarized with the expression 'increase in complexity' which implies increase in social differentiation, increase in economic interaction, etc. (Schallin 1997:19-20).

Schallin's view of urbanization once again highlights the inter-relatedness of the many variables that contribute to social complexity. Moreover, it also underscores the need to refine what is meant by "urbanization".

Modern conceptions of urbanization conjure images of large cities, densely packed apartment buildings and crowded streets. They imply high population density in a somewhat small, restricted space (L.A. excluded). But the first urban centers surely did not conform to this image. What, then, is urban in ancient terms?

According to Konsola (1986:9) "Urban is a society which is stratified, politically organized, based on a new and more complex division of labor." Konsola postulates that urbanization can be defined by multiple criteria – much as social complexity is. She cites these criteria as primarily socio-economic, geographic and demographic. Archaeologically, she sees these criteria as being defined architecturally.

Schallin sees the rise of urban centers as a consequence of cultural expressions. "...An increased demand for goods and services will eventually lead to craft specialization, social stratification and political organization...Political organization alone would not necessarily generate urbanization" (Schallin 1997:20). While both Konsola and Schallin may approach urbanization differently, they both agree that its ultimate rise is the end result of a variety of social and economic factors. They both view the development of urban centers as a probable consequence of regional economic activity. While Konsola prefers to identify urban centers from an architectural perspective, Schallin looks at the combination of architecture, craft specialization, social stratification and political organization in a settlement.

Despite differing approaches to the concept, the idea of a central place is key to most ideas of urbanization. However this central place need not be a town, it may also be a market place. In a central place model there may be a hierarchical arrangement between sites in a region i.e. the most important site is centrally located, surrounded by sites of less importance and complexity (Schallin 1997). Central places can be distinguished from urban centers, but this discussion is not predicated upon such a fine distinction. In the final analysis, urban centers, towns, and central places, develop in societies with intense economic activity and trade, which can generate social stratification. Whether these factors lead to, or are a result of, social complexity, is again, not central to this argument. The critical point is that they are all components of social complexity.

IV. Conclusion

These three facets of social complexity in particular were discussed as they are the ones that can be most readily seen archaeologically. Humans are dynamic and the ways in which they organize themselves can not be reduced to a simple linear model of social behavior. As such, this discussion of factors that mark socially complex societies is by no means exhaustive. Rather, it was intended to illustrate some of the variables that contribute to its rise. It was also meant to highlight that its definition is not clear, nor can the presence of any one of these variables mark with certainty a society as complex – or permit one to say that a certain period was characterized by social complexity.

As examination of just some of the facets of social complexity has shown (craft/economic specialization, political organization, social stratification and urbanization), the order in which factors contributing to social complexity arise is dependent upon one's point of view. But, an important thread which ties all viewpoints together is an agreement on which factors contribute to social complexity. In the end, the development of agriculture, an increase in population densities, the establishment of real permanent towns that function as marketplaces for the interchange of farm and craft production, and ensuing administrative areas of centralized power – are all factors that usher in social complexity and are a part of societies that are characterized by it.

This amalgamation of factors can be simplified to one succinct theory: the emergence of social complexity is usually associated with the decline of a subsistence economy and the rise of a political economy (Soderberg 2003). If this is the case, then data that directly addresses subsistence should hold the best potential for analyses of social complexity in ancient societies, such as faunal remains.

The animal bones may indicate a self-sufficient site in which production and consumption of food is contained intra-site with little or no trade. In this scenario, a less complex society would be hypothesized. Conversely, the faunal remains may suggest a producer or consumer site, in which case increased social complexity is suggested. Therefore, a key area this analysis will address is whether the presence of a subsistence economy is contrary to social complexity. In order to approach this issue, and others, one must understand faunal methodology. It is to this subject that this study now turns – animal bones and zooarchaeology.

CHAPTER 4

FROM FAUNAL METHODOLOGY TO SOCIAL COMPLEXITY

> *The study of animal bones to reconstruct animal economies from later prehistory may, at first sight, seem an irrelevant exercise. Without the thrill of Neolithic discovery, domestication and colonization to supply the analyst with the incentive of recording some new 'first' in the history of the domestic community, it may indeed appear a dull task* (Gamble 1982:161).

A keystone in anthropology is the study of social complexity. But what, exactly, is meant by this term? Anthropologists are primarily concerned with understanding humans - and humans must be understood within the context of their society. Society is comprised of complex social systems which are interrelated, forming the basis of social organization. Environment, subsistence, technology, contact and exchange, belief and cultural change, all comprise portions of societal organization. To this must be added categories of power, status, social hierarchy, culture, religious practices, economy, trade, ethnicity and wealth – and these are simplified categories that address just a few of the facets of a human civilizations. The term "social complexity" refers to the combination of each of these societal facets. For the purposes of this study, economic aspects of social complexity will be the focus – as it is the animal economy of Early Helladic society that is under scrutiny. Human-animal interactions are, however, at the core of every society, and by examining these interactions within the context of economics, we gain understanding of this vital aspect of social complexity.

Approaching social complexity via zooarchaeology is an often neglected pursuit. O'Connor (1996) notes the more complex a society becomes, the more zooarchaeology becomes sidelined in favor of other lines of evidence in archaeological interpretation. At a basic level this may be because simple questions of subsistence become over-looked in the interest of larger, more abstract questions. Yet, subsistence is integral to a society's economy. Faunal analysis is vital to the study of communal economies as it addresses the key area of subsistence. It is this economic specialization, particularly in regard to animals, on which this research focuses.

Understandably, the field of zooarchaeology has seen a growing emphasis on empirically oriented studies and taphonomy.[9] While important, these lines of inquiry tend to

---

[9] Lyman (1994:1) defines taphonomy after Efremov as "...the science of the laws of embedding or burial. More completely, it is the study of the transition, in all details, of organics from the biosphere into the lithosphere or geological record." However, he goes on to point out that these processes are not only non-human related, but

subsume other issues that are seen as less empirical, such as questions of social organization. Empirical studies can be used to inform and address less empirical questions. This approach enables the full potential of zooarchaeology to be reached. Lyman (1994) urges a return to more traditional definitions of taphonomy that take into account human activity as a taphonomic agent. Thus the fact that an assemblage was mediated by carnivores should be viewed as an opportunity to examine the relationship between carnivores and humans (Soderberg 2003). It need not be simply relegated to a taphonomic occurrence. We need interpretation along with analysis. Social factors must be considered along with taphonomic factors when studying a faunal assemblage. This analysis has such a methodology at its core, and will carry out interpretation of Early Helladic faunal assemblages as they apply to larger issues of social complexity.

How can we see social complexity zooarchaeologically?

> Zooarchaeology cannot produce a catalog of ethnic index fossils - the linkage between all types of social identities and material culture items is simply too complex for such a straightforward methodology. What can be done, however, is to specify the behavioral conditions under which social interaction took place and provide an understanding of the contextual constraints that structured innovations in animal symbology and their social meanings (Hesse and Wapnish 1997:238-9).

One of the easiest ways in which to view social complexity is through comparative analysis. Contrasting different chronological periods in a region can illustrate changes from more egalitarian societies (i.e. Neolithic Greece) to socially stratified, hierarchical societies (i.e. Mycenaean/Late Bronze Age Greece). The use of faunal analysis to address social complexity can also be woven into this comparative approach. Zooarchaeology can make use of the analysis of faunal studies from complex state societies by comparing them with assemblages from smaller, egalitarian settlements. In Early Helladic Greece, however, there are relatively few excavated and analyzed faunal assemblages, and simple comparative analyses are not always practical. Often, the bones from just one assemblage provide the corpus of material under analysis.

Animal bones provide information on their own about ancient subsistence practices, human-animal interactions (including domestication and animal husbandry), ancient environments, trade, inter- and intra-site economic relationships, and social hierarchies (Crabtree 1990; Kotjabopoulou et al. 2003; Redding 1991). These issues are all elements of

---

also human-related, and thus taphonomy needs to account for ALL processes that work on a bone or assemblage from the time of deposition to the time of recovery.

social complexity. This chapter will set out how faunal analysis can be used to addresses social complexity. It will first discuss standard faunal methodologies – or how exactly an assemblage of animal bones is analyzed and quantified once it is recovered. The remainder of the chapter will address methods of analyzing inter-site social and economic relationships. This chapter will provide the background and methodological underpinnings for chapters 7 and 8 – analysis of the Early Helladic faunal assemblages.

I. Faunal Methodology

The use of zooarchaeology to address conceptual questions, such as social complexity, begins with the excavation of bone in the field. This discussion will detail the steps faunal analysts take from the moment bones are recovered to their eventual interpretation.

Recovery:

Excavation methods differ from site to site and between geographic areas and chronological periods. Archaeological field methods often fall prey to the whim of individual field directors who shape the guiding questions surrounding an archaeological excavation. The unfortunate lack of methodological standardization can, and does, lead to divergent and varying data – and conclusions. Field methodology is often the product of a variety of factors, and will have a direct bearing on recovery procedures. These methods will depend upon the nature of the site, the size of the assemblage, the time allotted for excavation, as well as the importance placed on animal bones. Thus, a large site containing large-scale architecture, numerous small finds and significant amounts of pottery, may relegate animal bones to the bottom of the pile, so to speak, in importance.

Each of these factors will affect the amount of material that is recovered. If animal bones are not deemed important, or do not factor in core site analyses, recovery may be by hand only, with no sieving/screening carried out. If sieving is carried out, recovery can still be biased. The type and quantity of bone recovered will depend upon the size of mesh used in sieving, whether wet or dry sieving is done, and of course, the diligence and experience of the person carrying out the process. If collection of bone is mainly carried out by hand, smaller bones, such as those of the feet or smaller animals, may be missed. Larger bones, such as limb bones, teeth from large mammals, and larger mammals in general (cow, horse) will frequently be recovered. This has the potential to skew analyses of the relative numbers and types of species that were originally present in an assemblage. If wet/dry sieving is carried out, smaller bones and bone fragments, along with those of smaller animals, fish and birds, are more likely to be recovered, arguably providing a more accurate picture of the original faunal assemblage.

Studies have been conducted on the relative merits of different collection methods, and much ink has been used debating the correct way in which recovery should be carried out (Clason and Prummel 1977; Gamble and Bailey 1994; Meadow 1978; Payne 1972, 1975). Wet-sieving is often cited as the method lending itself to the greatest recovery of information, however other studies have shown, including the author's own, that often other factors come into play. The conditions and the soil from which the bones are recovered, as well as the diligence of those sieving and/or recovering the material, are often just as important as the use of a screen. At Helike, for example, the area under excavation lies under the water table. Thus, excavation occurs in wet sediment – mud, really. During the 2005 field season at Helike, I carried out an experiment with wet sieving the soil and found that while I did recover the occasional small bone fragment, the excavators were not missing much in the way of identifiable bone in the trench itself.

Standardizing field recovery methods, of course, is one of the roles of the faunal analyst in the excavation. Experience in the field at Helike indicated that bones were often broken in situ by the excavators, more often than not as a result of the field conditions and not due to careless excavation techniques. Since I am also in the field, I can make note of this, and this aids my eventual interpretation. However, too often bones are recovered without a site zooarchaeologist, only later to be handed over to someone to analyze and interpret – in the rarefied atmosphere of a lab. Faunal analysis can not be, and should not be, separate from work in the field. It is vital that the faunal analyst be out on site, examining context and working first hand with the material – in the trenches. This is one of the primary motivating factors for including Helike in this study. In addition, taphonomy plays a critical role in the analysis and interpretation of animal bone. Without first-hand field experience on site, many taphonomic factors may be over-looked or missed. This is an issue to which I will return shortly.

Identification:

After field collection, the next step in faunal analysis is preliminary sorting of the animal bone into identifiable and non-identifiable fragments. Once again, several factors contribute to this sorting, such as the experience of the analyst and availability and access to a reference collection. Fragmentation of the assemblage also plays a critical role in this process. This may be the result of taphonomic factors post- deposition, such as carnivore activity, soil ph, preservation, and excavation methods. These issues will be explored within the context of each of the faunal assemblages.

Once preliminary sorting is done, all identifiable fragments are sorted first by skeletal element, (i.e. femur, scapula) and then general category, such as small (cat), medium (dog, sheep), and large (cow, horse) mammal. Birds, reptiles, fish and rodents are often part of this initial sorting. Each specimen is then examined to ascertain whether the taxon can be identified, such as bovid, equid, canid, etc. Next, identification to the species level is attempted (donkey, wolf). Once the preliminary identification has been made, each bone is analyzed for sex, age, pathology, cut marks, burning, and other taphonomic factors such as carnivore gnawing and root etching. Naturally every bone will not yield every category of information. However, the more information one has, the greater the analytical potential and potential of the assemblage as a whole. If necessary, measurements are also taken on relevant bones, such as teeth or limbs, and aging data from tooth wear and epiphyseal fusion is also noted.

The depth and accuracy of identification will also be dependent upon the questions guiding the analysis, time allotted and size of the assemblage. In an ideal world, every bone would be measured, catalogued and quantified, however this is much easier in an assemblage of 50 bones than 50,000.

Taphonomy in greater detail:

> Sample representativeness is relative to some population which in turn is dictated by the research goal. The representativeness of a sample of faunal remains is controlled by the sample's taphonomic history, the sampling techniques used to collect the sample, *and* the research questions being asked of the sample (Lyman 1994:5).

Before proceeding to methods of quantification, taphonomic issues need to be carefully parsed out and their necessity understood. Taphonomy is the study of all processes - human and natural - that affect a bone after its deposition. Practically speaking, taphonomy is about changes to bones after death, before and after deposition. Lyman (1994) has gone to great lengths to address the importance of this study in zooarchaeology, and while recognition of its importance is growing, its application is often still dependent upon the nature of the site and the experience and motivation of the analyst and those in charge. In general, prehistorians rely heavily on taphonomy to understand questions surrounding early hominid hunting and scavenging. In more complex societies, such as Classical Greece, taphonomy is often little considered.

Taphonomy is a product of both cultural and natural forces acting on a bone assemblage - before, during, and after burial. An understanding of taphonomy provides information on a myriad factors affecting bone. Butchery patterns, impact by dogs or other scavengers, methods of refuse disposal and site maintenance by the former occupants of the area, coupled with the mechanical or chemical destruction of the ground, all impact bone preservation, and can lead to significant loss of bone data (Gamble and Bailey 1994). While excavation is itself an inherently destructive process, careful excavation coupled with sound methodology and an awareness of taphonomy, can mitigate loss of data and allow for the maximum amount of information to be gleaned from the bones themselves. Animal bones cannot be analyzed in the absence of taphonomy. It is perhaps the most significant area of analysis in zooarchaeology, and for that reason, will be addressed in greater detail. While specific taphonomic factors affecting each of the Early Helladic faunal assemblages under analysis will be addressed from within their respective analyses, this discussion will highlight those taphonomic factors most pertinent to this study. It will not recover ground already addressed by Lyman (1994).

Butchery:

An in-depth discussion of butchery is necessitated by the socio-economic focus of this research and the role that animal production and consumption plays in inter-site economic relationships. The eventual analysis of skeletal element frequencies and their variability between sites is intricately related to butchery practices and the analysis of cut marks on animal bone. In order to frame this eventual discussion, this chapter addresses butchery patterns in greater depth.

The past few decades have seen a great deal of research on bone surface modification caused by both humans and carnivores. In particular, discussion and research has been directed towards discerning the differences between human modified bone and other agents of bone surface modification, such as carnivores, trampling, root etching, etc. (Blumenshine et al.1996; Capaldo and Blumenshine 1994; Marean and Frey 1997; Pickering 2002; White 1992). Butchering can be defined "as the human reduction and modification of an animal carcass into consumable parts…consumable is broadly construed to mean all forms of use of carcass products, including but not restricted to consumption of products as food" (Lyman 1994:294). Extensive studies of hominid butchery patterns exist in the literature on early hominid behavior. Due to the large volume of work in this area, this discussion will not concentrate on the research on butchery patterns that relates to early hominid behavior (Binford 1981; Bunn et al. 1988; Pickering 2002; Shipman 1983; Shipman and Rose 1983). Rather, it will focus on those butchery patterns that are applicable to possible differences in the material inflicting the cut mark in question – a pertinent issue in Bronze Age Greece. The anatomical location of cut marks with the use of a particular type of tool will be examined, as will morphological, anatomical and quantifiable differences in cut marks inflicted during dismemberment, filleting, skinning and secondary butchery/food preparation.

Discussion of butchery practices must be prefaced with a note on cultural variability. Ethnographic data on modern hunter-gatherer groups indicates that butchery patterns vary widely between groups. Study of the butchery practices of the Aka (Gould 1969), the San (Yellen 1991), the Dassanetch (Gifford-Gonzalez 1989), and the Efe (Laden and Tappen) all indicate that butchery patterns are not universal, and may not always be correlated with anatomical joint articulation areas. Factors complicating the production of universal butchery patterns are variability in the tools used, the number of butchers, the purpose of the butchery (i.e. distribution or redistribution of meat), and cultural variability. Given the hindrances to producing a template universal to all butchery patterns, a few general conclusions about the traces butchery may leave on animal bone can still be drawn. It is useful to examine the various stages in butchery and the cut marks inflicted during those stages, in order to offer a template from which to drawing further conclusions.

Binford (1981) is one of the few researchers to address the differences in cut marks resulting from the material used in butchery. He conducted a comparison of fauna butchered by the Nunamiut Eskimo with metal tools with data gathered from the Mousterian site of Combe Grenal (butchery by stone tools). His study assumes a butchering sequence that begins with skinning, followed by dismemberment, filleting (for consumption or storage – often involving further dismemberment) and marrow consumption. Binford's study drew several conclusions on the anatomical placement of various types of butchery marks that are summarized here.

Skinning: According to Binford (1981), there are relatively few places on faunal anatomy that bring the butcher in direct contact with bone. However, the two places where direct contact is most likely to occur are the lower limbs and the head. Skinning activity usually leaves cuts that appear to encircle the shafts of lower limb bones, such as on the lower tibia, the shaft of the metatarsal, and the phalanges. Encircling cuts (indicating skinning marks) have also been noted on the distal shaft of the radio-cubitus. This encircling cut is generally one of the first to be made, and subsequent cuts may originate here and extend down the medial face of the limb. If the head is skinned, cuts may be visible around the base of the antlers, horns, around the ears, mouth and chin area of the mandible. Binford (1981) also noted that variability might exist in skinning marks depending upon whether skinning was done as a state of butchery or for skins.

Dismemberment: Dismemberment, or primary butchery, may leave some of the most distinctive cut marks. Dismemberment usually consists of skeletal disarticulation and so cut marks are often associated with anatomical points of articulation (Binford 1981).

Dismemberment marks are commonly found in association with removal of the head, sometimes acting on the articulation between the atlas and axis vertebrae; removal of antlers/horns may also take place, in which case the skull case is broken into and fragments of skull may be removed. Cut marks may appear from disarticulation of the mandible along the masseter muscle, and around the medial margins of the third and fourth premolars if the tongue is removed. Butchery of additional vertebrae is usually done during filleting or secondary butchery, and is largely dependent upon the size of the animal. Primary butchery marks may also appear on the ribs and sternum, and just anterior and posterior of the acetabulum on the pelvis (for disarticulation of hind limbs).

In larger animals the sacrum may be disarticulated from the pelvis, resulting in longitudinal cut marks down the iliac wings. In many cases, especially with smaller animals, removal of the lower limbs is one of the few actions taken in primary butchery. Thus, cut marks are concentrated on the ball of the femoral head, femoral neck, as well as on both femoral trochanters (with marks being more common on the greater trochanter). Generally, the proximal tibia remains articulated with the distal femur prior to secondary butchery. In situations involving larger animals and attendant field transport issues, femoral-tibial articulation may be carried out, in which case marks are often found on the patellar surface and on the posterior surface just above the lateral and medial condyles (Binford 1981). Similar patterns are noted in dismemberment of the front limbs. Some primary butchery techniques leave cut marks on the scapula that encircle the glenoid cavity and sometimes the scapular neck. Often, however, the scapula remains articulated with the rest of the forelimb during primary butchery. Thus, higher frequencies of scapular cut marks are associated with secondary butchery. There are also few primary butchery marks on the proximal humerus.

The distal humerus bears more numerous and consistent cut marks, commonly across the medial and anterior faces.

Filleting: Filleting is a process that may occur as primary butchery (often related to the size of the animal being butchered as well as transport issues), or it may occur after the initial disarticulation and distribution of meat. In such cases, the filleting may be carried out by women and others not involved in the initial stage of butchery. Often, filleting is thought of as a stage of food processing in which the meat is prepared for consumption or storage. Secondary butchery often involves further segmentation of the parts cut up during primary butchery, usually in preparation for cooking.

Thus, marks from secondary butchery may be most common on lumbar vertebrae, the pelvis and the limb bones (Binford 1981). In his study of the Nunamiut, Binford emphasized that filleting marks may leave longitudinally oriented marks in respect to the bones on which they appear. He stressed that the very act of filleting dictates this pattern. In association with these longitudinal marks, shorter, more obliquely oriented marks appear along the posterior or anterior surfaces of long bone diaphyses. Such marks will be clustered when the shape of the bone is irregular and where there are numerous muscle insertions. Filleting marks would be expected to occur most frequently at the articulations of the femur and tibia/fibula, the humerus, radius and ulna. This assertion is based on Binford's observation that the limbs commonly remain articulated during primary butchery and are frequently disarticulated after division and secondary practices. Following filleting and secondary butchery, the breakage of long bones for marrow extraction may occur, a process that has a well-developed body of literature in its own right (Binford 1981; Pickering 2002; Villa and Mahieu 1991; White 1992).

Additional factors affect the appearance of cut marks on bone. One such factor is the protective role of the periosteum. Soft tissues, such as the periosteum, have an important ability to shield bones from being marked by either stone or metal tools, and may help explain low percentages/absences of cut marks on bones of known butchered animals (Shipman and Rose 1983). Additionally, study examining the role of the load force and cut mark morphology concludes the depth/width of a tool mark is directly related to the amount of force applied (Walker and Long 1977).

Examination of the comparative butchery data analyzed by Binford (1981) suggests differences in cut marks between assemblages butchered by stone tools and assemblages butchered by metal tools. As the frequency of use of metal tools in Early Bronze Age Greece is an area about which little is known, Binford's data is certainly relevant. In brief, fauna from the Mousterian site Combe Grenal (butchery by stone tools) contain a high frequency of butchery marks on the mandible, antler bases, scapula, acetabulum and the distal humerus. Fauna butchered by modern Nunamiut with metal tools contain high frequencies of cut marks on the phalanges, the distal humeri and femora, and the carpals. Comparison of these data do suggest differences in the anatomical placement of cut mark frequencies, however it is unclear whether these differences are attributable to material differences in tools. Ethnographic studies of modern hunter-gatherers reveal variability among different ethnic groups, suggesting that cut mark placement may be reflective of cultural differences and not necessarily of tool variability. Taphonomic factors can also affect these data. It is therefore not useful to draw conclusions concerning differential anatomical placement of cut marks resulting from stone and metal tools alone. Additional data is needed to thoroughly address this question.

Bone density:

Preservation of animal bone in archaeological contexts may be mediated by the relative densities of the bones themselves. This is a factor with direct bearing on the analysis of the types of species noted in an assemblage, as well as with the different parts of the animals identified. In brief, some skeletal elements have greater bone density than others. "...The probability that a skeletal part will survive the rigors of various taphonomic processes is at least partially a function of that part's structural density" (Lyman 1994:235). Thus, long bone shafts (diaphyses), having a higher concentration of periosteal or compact bone, tend to be denser, and thus stronger, than long bone ends (epiphyses), which have a higher concentration of cancellous or spongy bone. This translates to better preservation of the middle portions of long bones and poorer preservation of distal ends. Identification of species is brought to bear on this factor, as it is often the distal portions of a long bone that contain morphological traits most useful in species identification. This is also an especially pertinent factor if the assemblage in question was subjected to carnivore activity. My own experimental study with a 30 kilogram German shepherd reduced the femur of a goat to a few long bone shafts – all but unidentifiable. The same dog inflicted little damage to a cow femur, save gnaw marks on the distal ends. This actualistic study confirms the theory that as bone density increases, the frequency of identifiable skeletal parts increases.

Bone density is also brought to bear on the type of animal in question. Larger animals, such as cows, tend to have denser, more solid bones, than smaller animals, such as cats. This may affect the bones that are actually preserved in an assemblage, and may ultimately paint an inaccurate picture of the relative proportion of different animals present at a particular site. Several studies have been carried out on the actual bone densities of different skeletal elements both intra- and inter-species, and will not be reiterated here (Behrensmeyer 1975; Brain 1969; Ioannidou 2003; Lyman 1994, Lam et al. 1999, Lam and Pearson 2004; Novecosky and Popkin 2005). Bone density is an area to which I will return from within discussion of the specific Early Helladic faunal assemblages under analysis.

As previously mentioned, taphonomic analyses are multi-faceted. This discussion was meant to serve as an introduction into some of the most pertinent aspects of taphonomy as related to Early Helladic faunal assemblages under analysis in this research. Taphonomic issues will be further discussed in relation to each faunal assemblage. Attention now must be directed at the analyses that directly result from these taphonomic factors - amalgamation, sample size and methods of quantification.

II. Quantification – or the Fascinating World of MNE's, MNI's, NISP's and MAU's

The endpoint of zooarchaeological analyses is fundamentally to draw conclusions about the society who utilized the animals under study. In order to arrive at this desired interpretative endpoint, we need to have some way of counting up the number of bones in an assemblage, and then drawing out meaning from them. Several methods exist which facilitate the analysis of sample size and amalgamation in faunal assemblages. These methods include the most common quantitative units of Number of Identified Specimens Per taxon (NISP) and Minimum Number of Individuals (MNI), to more recent analyses that calculate the Minimum Number of Elements (MNE) present and Minimum Number of Anatomical Units (MAU). Each method of quantification has its relative merits, and the choice of which to calculate will often be governed by one's questions, sample size and the nature of the assemblage. Before delving into the details of each method, a word must be said about sample size.

The larger the sample size and greater proportion of the site excavated, the more accurate picture of the original faunal assemblage one is able to reconstruct. This is in an ideal world, however, and in areas such as Early Helladic Greece, in which few faunal analyses have been carried out, it is necessary to find a starting point. This means that while a relatively small sample size, such as Helike, may not support the kinds of conclusions of a larger assemblage, it can still serve as a starting point. At the present time little can be done about the small quantity of animal bone already recovered in Greece. It is far better to begin discourse that illustrates the potential of faunal analysis so that in the future researchers will be more inclined to conduct analysis. Sample sizes will increase with recognition of zooarchaeology's potential. Now – on to the numbers!

Number of Identified Specimens (NISP):

Along with MNI, NISP is the quantitative unit most frequently encountered in the published literature. It is an observational unit that simply counts the number of identified bones in a faunal assemblage. So, if 30 bone fragments were identified the NISP is 30. NISP can have a meaning identified to any level – it may be to taxon, species or even family. Thus, a NISP of 30 may mean 30 fragments from a medium sized mammal, or it may be more specific, and mean 15 identified pig fragments, 10 sheep, three goat and two dog. One of the problems with NISP based species ratios is that they fail to take into account skeletal elements that come from the same animal. Thus, a calculated pig NISP of 100 may reflect the bones from just one individual. Calculating MNI avoids this problem. A second problem with relying solely on NISP is that the number is greatly affected by the degree of fragmentation of the assemblage.

Minimum Number of Individuals (MNI):

Unlike NISP, MNI is a derived unit. It is "the minimum number of individual animals necessary to account for some analytically specified set of identified faunal specimens" (Lyman 1994:100). It is a derived measurement that may not take into account variables such as age, sex, or size. If, for example, there are three left pig femurs in an assemblage and one right femur, then the MNI for pigs is three. That is, at least three individuals would have had to be present in the original assemblage to account for the three femurs. However, what if the right femur is from a juvenile and the left femurs are all from adults? If this is the case, at least four individuals were originally present, providing an MNI of four. In order to avoid this confusion, analysts state whether they are "maximizing MNIs" by accounting for variables such as size, age, and sex of skeletal elements. In addition, measurements of MNI need not be assigned to individual taxa; they may just be assigned to element. For example, an assemblage may have five right scapulas that can not be assigned to individual species.

A simplified way to calculate MNI traditionally used is to determine a distinct MNI for each skeletal element – this would take into account left and right sides, and how many times that element occurred in a given individual. Thus, if there are four canid left femurs and three right femurs, there is an MNI of four canids based on femora. However, there may be two left canid scapula and one right, and thus the MNI for canids based on scapula is two. Once again, however, this method is blind to age, sex and other variation between individuals. One of the most prevalent problems with conclusions based on MNI calculations is the unpredictable response to the problem of aggregation. A solution to this issue, and one adopted in this dissertation, is to present both MNI and NISP calculations, and when presenting MNI data to do so by skeletal element.

Minimum Number of Elements (MNE):

While MNI and NISP measure the frequencies of taxa, MNE measures the frequencies of portions of skeletons of individual taxa – an analytical method tied to the more recent emphasis on taphonomy (Lyman 1994). MNE signifies the number of a particular skeletal element necessary to account for a portion of a skeletal element. Perhaps the easiest way to understand MNE by looking at two ways it can be derived. The first method involves measuring something like the complete circumference represented by a long-bone shaft fragment, and then summing those proportions for each portion of a skeletal element (Lyman 1994:103; Marean and Spencer 1991:649-650). It can also be derived by simply calculating the portion of the bone represented by the fragment and then summing the fragments to arrive at an MNE for each skeletal portion. Thus, if one fragment of bone represents one half of a proximal humerus and another fragment represents one half of a distal humerus, an MNE of one humerus would be calculated. Additional ways to calculate MNE have also been attempted (Bunn and Kroll 1986), and as with the above methods, all have their strengths and weaknesses. As Lyman (1994) suggests, the best way in which to navigate this web of conflicting methodologies is to be explicit about the way in which the calculation was performed. This allows for other researchers to more accurately compare two assemblages.

One quantification method that may be particularly useful in eventual interpretation of the faunal data is the ratio between NISP and MNE. Dividing NISP by MNE can provide an idea as to the degree of fragmentation in an assemblage. This can examine fragmentation of different skeletal elements, but also different taxa. It can provide clues as to taphonomic processes that may have affected the assemblage as a whole, such as carnivore activity or extraction of marrow by humans.

Minimum Animal Units (MAU):

An acronym standing for the *M*inimum *N*umber of *A*nimal *U*nits necessary to account for the specimens in a collection, MAU's are a standardized variation on MNI calculations. Developed by Binford (1984), MAU's are calculated by dividing the MNE for each anatomical unit (i.e. a proximal femur) by the frequency of their occurrence in an animal (in this case, two). This method aims to analyze survivorship of different skeletal parts, and reflects how many of each of the various portions of carcasses are represented.

The relative merit of each method of quantification has been debated and responded to at length by scholars. The intent here is not to rehash these debates, but instead to note the types of quantification that are possible in faunal analysis. This discussion was also intended to illustrate the potential for interpretive variability depending upon the type of quantification method used (see *Table 7.2* for a comparison of NISP:MNE:MNI). The ensuing chapters' analyses on the Early Helladic material will make use of most of these methods, and justification and discussion of each will take place within those contexts. With faunal methodology laid out, discussion can turn to practical application of faunal data to the wider issue of social complexity.

III. Inter-Site Socio-Economic Relationships

> By examining certain aspects of the faunal sample, we can discover the nature of the animal production strategy thereby allowing us to gauge the outward or inward focus of an important facet of the economy and its relation to the larger regional pattern (Wapnish and Hesse 1988:83).

Fieldwork and the ensuing faunal analyses involving identification and quantification in an assemblage are too often the point at which faunal analyses have stopped – at least in Early Helladic Greek contexts. This study advocates application and interpretation as the next indispensable steps in faunal analysis. Using the raw data, variation between faunal assemblages at different archaeological settlements facilitates the reconstruction of several aspects of social organization. At a general level, social differentiation can be addressed by using the faunal data to ascertain whether social stratification exists in the early Greek Bronze Age. This can be accomplished through examination of the types of fauna exploited, as well as the anatomical elements present and their relative frequencies (Ijzereef 1988). In addition economic complexity can be addressed by reconstructing inter-site economic relationships involving the exchange, redistribution and trade of animals and animal products (Clark 1987).

Examining age profiles (kill patterns) and sex ratios of different species in assemblages facilitates analyses as to the use of particular animals. Abundant research exists, particularly in Europe and the Near East, concerning harvest profiles and their relationship to different economic strategies (Crabtree 1990). This research has yet to be accomplished in Early Bronze Age Greece.[10]

Producer vs. consumer sites and socio-economic complexity:

One way in which to approach differences between archaeological sites is to examine the possibility of the existence of a specialized animal economy (Soderberg 2003). Do some settlements serve as production centers for animal products (producer sites) while others function as consumers for these animal products, trading for other goods? In other words, is there a recognizable economy centered upon animals that is visible in the faunal assemblage, and if so, how can it be identified and what does it mean?

Zeder (1991) has conducted extensive work on animal economies and the provisioning of cities. She examined animal remains in urban and rural contexts in the Near East in order to ascertain if some settlements were acting as centers of animal production while others acted as centers of consumption. The existence of differentiation between settlements – some of which function as production centers while others function as consumer centers, speaks to the larger issue of economic specialization and economic complexity, wherein social and economic relationships are maintained between sites based upon trade and resource acquisition. This economic specialization suggests, at the very least, the existence of societies organized into chiefdoms (Pullen 2003), and more likely, the existence of social complexity within the spatio-temporal context under study.

At the center of the producer/consumer dichotomy and social complexity, is the much broader concept of human/environment interaction – an environment of which animals are a part. Flannery's Systems Theory (1968, 1971) treats humans and the environment as "a single complex system, composed of many subsystems which mutually influenced each other" (Flannery 1971:345). Renfrew was the first Aegean prehistorian to make use of general Systems Theory, specifically to elucidate the development of complex societies in the region (Renfrew 1972). "Each decision had ramifications beyond the realm of subsistence, so change spread through all parts of the cultural system. The ecological perspective stresses this dynamic mutual relationship between culture and environment" (Kardulias 1992:425).

---

[10] Research does exist on the role of secondary animal products on the palatial economies of Late Bronze Age Mycenaean Greece, but this research does not reach back into the Early Bronze Age (Halstead 1981, 2003).

This ecological perspective then has the ability to change, alter and restructure societies, ultimately influencing social complexity.

A common feature of all complex societies is a degree of specialization and differentiation in production. The concept of specialization between sites involves not only economic differentiation (i.e. division of labor), but also economic individuation, in which certain persons spend a significant portion of their time in particular non-subsistence activities that do not engage others (Kardulias 1992). Thus, it is clear that an economy containing specialized animal production also contains specialized craft and resource production. Again, these additional specialties suggest a complex social system. As Evans has noted, specialized production requires that "these specialists cannot participate in some or all of the basic community subsistence goods through exchange for their craft products" (Evans 1978:115). Thus, if producer and consumer sites can be identified in Early Helladic Greece on the basis of faunal analysis and inter-site comparative studies (such as Helike, Lerna, Tsoungiza and Tiryns), then zooarchaeology may support the idea espoused by Pullen (2003) that Early Helladic Greece sees the beginnings of the social complexity that marks later periods.

Crabtree (1990) has examined how animal bones have been used in the study of complex societies and identifies the kinds of zooarchaeological methods and techniques critical to these studies. She identifies two basic questions: How are animal products used in trade and exchange? And can faunal remains be used in the study of complex societies? Crabtree stresses that processes of state formation or urban development have been the focus of many archaeological studies of the evolution of complex societies, and these processes are critical to the emergence of cultural complexity (Crabtree 1990:156). Zeder (1991, 2003) in particular has shown that faunal remains can contribute to our understanding of urban processes. However, Crabtree points out that a society need not be an urban state to possess social complexity.

> From a theoretical standpoint, the line between a complex chiefdom and an incipient state is a fine one. The problem is particularly difficult when one is trying to distinguish chiefdoms from states based on purely archaeological criteria. Complex societies (i.e., both chiefdoms and states) can be distinguished archaeologically from simpler hunting-and-gathering and agricultural communities based on the presence of such features as well-defined social hierarchies, regular and extensive long-distance trade networks, craft specialization, and complex multitiered settlement hierarchies. It therefore seems reasonable to treat all complex societies as a single group...(Crabtree 1990:156).

Returning to Pullen's idea (2003) that Early Helladic Greece is a society loosely organized into chiefdoms, this distinction becomes all the more salient. As Chapter 2 has shown, Early Helladic Greece likely possessed all the trappings of a chiefdom (and thus a complex society), such as trade networks, craft specialization and settlement hierarchies based upon size differentiation. With these ideas in mind, it becomes necessary to examine just how economic specialization and animal economies can be recognized in the faunal record.

How can we identify producer and consumer sites using faunal analysis?

An issue for complex societies in regard to economic specialization is the provisioning of non-food producing specialists with food. Faunal analysis provides one way in which archaeological data can be used to study this provisioning process. Consumers can be provided with food via large-scale distribution, market exchange and through individual exchanges between producers and consumers (Crabtree 1990). In Early Helladic Greece, it is most likely the latter that comes to bear most significantly on the economic system. Thus to reconstruct social complexity via zooarchaeology from an economic perspective, trade needs to be understood. Trade can be examined by the ways in which meat and other animal products are exchanged between producers and consumers. Following Crabtree (1990) we can ask the following questions:

- Were meat and other animal products provided by specialist producers?
- Can we identify producer and consumer sites, or are they one and the same?
- What types of meat/animal products were available to consumers?
- How are animal products made available?

Each of these questions revolving around trade also centers on the identification of producer and consumer sites zooarchaeologically. The redistribution of animal products should produce some degree of patterning of faunal assemblages at producer and consumer sites. This patterning can be identified with a few key zooarchaeological analyses: 1. inter-site species comparisons, 2. inter-site age/slaughter profiles, and 3. comparison of skeletal element frequencies between sites/assemblages.

Methods - Inter site species comparisons and skeletal element frequency:

Faunal assemblages may differ in the types of species present at settlements undertaking the raising of livestock compared with settlements consuming the animal products. Examining the range of species present in a faunal assemblage provides an insight into the overall range of animals exploited, and ultimately addresses variability in diet and dietary patterns. In an ideal environment, a faunal assemblage would accurately reflect the animals that were indeed present. However, archaeological assemblages are all too often far from ideal. The range of animals documented is a function of overall sample size, and species ratios are often based on quantitative studies that attempt to reconstruct what the original animal population may have been. Each method of quantification has been subject to critical reviews, and often the actual method used will reflect the questions asked by the faunal analyst and the nature of the faunal assemblage. No single technique has been proven to be 100 percent accurate.

Criticism of quantification aside, in general there should be recognizable patterns in faunal assemblages suggestive of each type of site. Zeder's work in the Near East has shown that as direct contact between herders and consumers declined, non-producing consumers were increasingly provisioned with a narrower, limited range of animal species (Zeder 1991)). Thus faunal assemblages at consumer sites should yield evidence for the remains of a relatively small range of species. In urban contexts throughout the Mediterranean, faunal assemblages at consumer sites tend to be dominated by *ovicaprids* (sheep and goat). Smaller settlement contexts that rely heavily on local animal production and hunting yield evidence for increased diversity in species – with pig comprising a significantly larger portion of the assemblage, sheep/goat declining, and the occasional wild fauna present. This is an idea to which a great deal of attention will be turned in the ensuing chapters.

Crabtree (1990) argues that we should not necessarily make the assumption that faunal assemblages in all areas of the world and all time periods will reflect this pattern. While she cites examples from later Medieval European and North American contexts, what little work has been done in Greece appears to corroborate Zeder's findings. In Late Bronze Age contexts, "urban" assemblages, or those from higher density occupations in the environs of palatial contexts within fortification walls, show a heavy dependence on sheep and goat. In most Late Bronze Age contexts in Greece, sheep and goat are the heaviest exploited animals (see especially Tiryns, von den Driesch and Boessneck 1990). Crabtree points out that while there is a significant body of evidence on assemblages from larger sites, not enough work has been conducted in smaller settlements to test this assertion. Faunal analysis at Helike will provide this small settlement data and enable this study to test both Zeder and Crabtree's ideas, forming an integral part of the socio-economic analyses that will be undertaken.

Methods – Age/slaughter profiles:

Examination of the relative ages at which individual domestic animals were slaughtered, taken in conjunction with the range of species present, also offers insight into inter-site economic systems. Many animals, such as sheep, goat and cattle, are used not only for their primary products (meat), but also for secondary products, such as milk, cheese and wool. In order for their role within a faunal assemblage to be accurately ascertained, it is necessary to examine the age at death. The age at death is primarily determined based upon two methods: tooth wear/eruption data and epiphyseal fusion data from long bones. Various studies have been made for the major animal domesticates that have formulated general age categories for the fusion of different skeletal elements (Silver 1969). The bones in an archaeological assemblage can be examined for these skeletal markers, and an age estimate can often be given for an individual. One hindrance to aging on fusion data is that once bone growth ceases (often in early adulthood), these data are no longer relevant. Thus, age estimates based upon dentition are probably the more common of the two analyses, as teeth tend to experience some of the best preservation, as well as change continually throughout an individual's lifetime. Established tooth wear stages and eruption data are available for most of the major domesticates (Halstead 1985, Grant 1982). Studies have concentrated on cattle (Grigson 1982), sheep/goat (Deniz and Payne 1982; Payne 1973) and pig (Bull and Payne 1982) in particular. Age data derived from the combination of dentition and epiphyseal fusion (Silver 1969) provides an excellent method for determining the probable function of an animal assemblage.

Domestic animals are raised for a variety of reasons, from use as draft animals to production of primary and secondary products. Some animal products may be collected at intervals (wool) while others collected only once – at slaughter. For example, younger individuals tend to have served a primary function as meat producers. If the majority of an assemblage is comprised of younger individuals, the conclusion can be reached that they were exploited for their primary product – meat. If however, the assemblage in dominated by older individuals that offer secondary products, we can begin to ask different questions of the animal economy – and that is a possible trade/specialization in secondary products (Chang 1994; Clark 1987; Crabtree 1990; Greenfield 2003, 2005; Halstead 1981, 1996, 2003; Sherratt 1981; Zeder 1991, 2003).

Specialized economies based upon secondary products played a key role in the development of social complexity. An absence of written records, such as in Early Helladic

Greece for example, makes this an area more difficult to approach. However, we can still hypothesize this took place by looking at mortality patterns and species ratios. An assemblage dominated by a high proportion of adult sheep, possibly including a high proportion of wethers (castrated males) might indicate sheep raised in the production of wool. During the Late Helladic period, Linear B evidence tells of the role of secondary products and animals in the Mycenaean palatial economy. In the absence of the written record, evidence must come purely from the bones themselves.

As with any analysis, there are problems with aging methods. A problem highlighted by Wapnish and Hesse is that of interdependence. That is, one age class of animals in an assemblage may be represented by a number of bones from each individual, while another age class may be represented by just one bone per dead animal. If the two age classes are analyzed together than the contribution by the first age class will be overestimated. They suggest this problem can be overcome by the use of MNI's. Thus, the frequency of the most common element type would provide the basis for aging data, which would then be used in the calculation of harvest profiles. This would provide a more accurate picture of the original animal population.

Wapnish and Hesse (1988:84) have proposed three models for the production and consumption of domestic animal resources based upon their work in the Near East that have direct bearing on harvest profiles. Rather than drawing a strict line between producer and consumer sites, they allow for the addition of a third type of site – a self-contained economy. In a self-contained economy animals will be produced and consumed locally, and thus harvest profiles will include all age classes. In a consuming economy locally raised animals are supplemented by acquired individuals (either through trade or exchange) and harvest profiles should include an abundance of market-aged animals (1-2 year age range, depending upon the species) and relatively few individuals of reproductive age. Producing economies should yield aging data on either extreme – very young and very old individuals, as well as those older individuals culled from breeding stock. While no one site will in all likelihood satisfy these profiles exactly, these general trends provide working models on which conclusions can be drawn.

Methods: Skeletal element frequencies (body part distribution):

Analyses of skeletal element frequencies center on the portions of an animal that are most valuable as food – or the prime meat bearing bones. For example, pigs are an animal that typically serve a sole function as a meat producers. This means that, unlike sheep, goat and cattle, they offer no secondary products. Therefore, the presence of pig remains at a site can be interpreted to have served the primary role of sustenance. In order to ascertain, however, where pigs were raised, skeletal elements need to be examined. A predominance of head and foot bones (low utility elements) can be variously interpreted, however. Such a distribution may indicate the settlement in question was engaged in rearing the animal. If the prime meat-bearing bones, such as upper limbs (humerus and femur) are also present, the conclusion can be drawn that the animals were being raised and consumed in the same place. If, however, one finds an absence of prime meat-bearing bones, a tenable hypothesis may be that the animal was being raised on site, butchered, and the higher valued portions involved in trade or exchange with another site – indicating an inter-site economic relationship. It is also possible, however, that live animals were traded or exchanged at the consumer site itself. In this case, the skeletal element frequencies would mirror a self-sustained site. This is where analysis of butchery patterns and cut marks becomes vital. If cut marks are present on some of the elements, they can be analyzed as having resulted from either primary or secondary butchery (Binford 1981). The placement and distribution of cut marks can then be viewed against skeletal element frequencies, and hypotheses drawn as to the place of butchery of the animal in question. We can also turn to the relative distribution of skeletal elements on the site itself. Perhaps one area is being used for butchery (yielding a lot of head and feet), while areas with households yield the higher valued parts of the animal. These data could then be used to approach questions regarding intra-settlement social differentiation, with higher utility elements appearing in wealthier contexts and vice versa.

A complicating factor in analyses of skeletal element abundances are the occurrence of reverse utility profiles (disproportionately high frequency of body parts with low food value, such as cranial elements) which may be caused by preservation bias (low economic utility elements tend to have the highest bone density, and therefore increased probability of survivorship). Such profiles are well documented in artiodactyls, and have direct bearing on pig element survivorship. These profiles, however, may also be the product of human agency. High economic utility elements, such as the femur and humerus, are more susceptible to fragmentation from human agent (marrow extraction) and carnivore activity. Reverse utility profiles can thus be misleading, prompting erroneous economic hypotheses to be drawn. One way in which to mitigate against false conclusions produced by reverse utility profiles is to account for bone density and its role in assemblage formation. However, one must also account for additional variables. Numerous studies have suggested that reverse utility profiles may be more reflective of analytical biases than human behavior (Lyman 1992; Marean and Frey 1997; Marean and Kim 1998). These analytical biases are a product of the quantification methods used, and can be mitigated by careful use of a variety of measurement techniques. Quantification methods that shed light on the degree of fragmentation of the assemblage, such as NISP/MNE ratios, can help ascertain whether high utility elements actually are present, but overlooked due to fragmentation imparted by marrow extraction, cooking or carnivore activity. Accounting for shaft fragments in analyses provides insight into this potentially sticky area, and can help recreate the actual skeletal element frequencies.

A second potentially complicating factor in the analysis of skeletal element frequencies concerns the problem of equifinality, in this case, the variability in the number of bones between species. For example, pigs have twice as many foot bones as cattle. Thus when comparing the relative frequencies of foot elements between these two species using NISP, pigs may appear to have more foot bones than cattle in an assemblage. This can lead to erroneous assumptions concerning differential use between species, as well as inflate the number of pigs in relationship to other species with fewer foot bones. Problems arising from this factor can be mitigated through careful quantitative analyses, however. Comparing inter-species data using either MAU or MNI would alleviate this issue, as use of either method accounts for the variability between species within a body part (e.g. pig foot NISP=12, but MNI=2 vs. cow foot NISP=5, but MNI=3). If quantification methods which account for the total number of bones per species within a body part are used, this potential complication can be avoided. However, also of relevance to this issue is differential survivorship of certain elements, due either to taphonomic factors, such as bone density, or other factors, such as transport and carnivore activity. Use of the previously discussed reverse utility curves, coupled with analysis of possible bone density mediated attrition and discerning use of quantification methods, are all ways in which to mitigate potential biases as much as possible.

The discernment of producer and consumer sites from analysis of faunal assemblages is dependent on a large variety of factors. As already discussed, taphonomy plays a large role in the preservation and eventual excavation of different skeletal elements – and species. Recovery methods play an equally important role. Thus, the best way in which to approach a faunal assemblage is to take all analyses together to create a full picture of the economy in which animals played a key role. With respect to the identification of producer and consumer sites from faunal analysis, methodologies must incorporate all possible areas of evidence.

Thus, the species present, mortality profiles and skeletal element frequencies must all be analyzed together in order to arrive at tenable hypotheses concerning the animal economy – and ultimately social complexity.

IV. Conclusion

This chapter has shown that there is a fundamental relationship between humans and animals that is at the heart of economic complexity, and ultimately, social complexity, in ancient societies. It has illustrated the absolute relevance of faunal remains to study of ancient settlements. Through a discussion of standard faunal methodologies it has laid the groundwork for a critical analysis of Early Helladic faunal assemblages at Helike, Tsoungiza, Lerna and Tiryns. Archaeological evidence suggests that Early Helladic Greece may have been organized into loose chiefdoms, and thus possessed the beginnings of social complexity. There is evidence for regional variation, with similarities and differences derived from pragmatic material concerns. A model of human behavior based on rational decision-making underlies this study, and suggests that individuals act universally to increase benefits and minimize costs in any actions they undertake. Animals can be this buffer and certainly played a vital role in human actions through time.

Early Helladic Greece clearly had some type of redistributive economic system (Renfew 1972) or a system of social storage (Halstead 1981) in which animals played a key role – but what was the nature of this system and what does it indicate about Early Helladic society? What remains is to test this theory of a socially complex Early Helladic period through a critical examination of the animal bones themselves. Part II of this study will present the faunal data from each site which will then be used to examine inter site socio-economic relationships and eventually reconstruct social complexity in spatio-temporal setting. However, before this can be accomplished, one species in particular, pigs, deserve a detailed look in order to build a case for their utility as predictors of economic aspects of social complexity in ancient settlements.

CHAPTER 5

WHY PIGS?

Lisa: *No I can't! I can't eat any of them!*
Homer: *Wait a minute, wait a minute, wait a minute. Lisa honey, are you saying you're **never** going to eat any animal again? What about bacon?*
Lisa: *No.*
Homer: *Ham?*
Lisa: *No.*
Homer: *Pork chops?*
Lisa: *Dad! Those all come from the same animal!*
Homer: *[Chuckles] Yeah, right Lisa. A wonderful, magical animal.*[11]

The pig (*Sus scrofa*) has played a significant role in the history of humanity as a valuable food resource. Pigs are found at nearly every archaeological site across the Mediterranean and temperate areas of Europe diachronically. Pigs have played a vital role in human subsistence since the Paleolithic, reaching an apex during the Bronze Age, and maintaining a steady presence to modern times. During the Neolithic and Bronze Ages in some areas of Europe and the Mediterranean, pigs account for nearly 40 percent of the total food species recovered, and while their relative numbers drop in later historic periods, they consistently account for a high percentage of the species exploited.[12]

Given their importance as a resource, the paucity of study accorded the pig in socio-economic contexts at archaeological sites in Greece is astounding. With the exception of work in the Near East and the Levant (Diener and Robkin 1978; Flannery 1982; Grigson 1987; Hesse 1995; Hesse and Wapnish 1997, 1998; Hongo and Meadow 1998; Redding 1991; Zeder 1996, 1998) pigs often receive little attention beyond simple discussions of relative numbers in archaeological reports.

The aim of this chapter, therefore, is to explore the history and importance of this widely utilized animal by examining its phylogenetic history, its domestication, and finally its presence in archaeological contexts. This discussion will also address the value of pigs as a resource and lay out the model used by this study for understanding economic aspects of social complexity via pig exploitation. It will also address hypotheses for pig utilization in the zooarchaeological literature, which will then be applied to Early Helladic Greek contexts.

---

[11] *The Simpson's* Episode 3F03
[12] In the Levant and Near East see Hesse (1995) for data on Neolithic Merimde in Egypt (39% pig), Predynastic Heirakonpolis in Egypt (41% pig), New Kingdom Amarna in Egypt (47%), EB Hassek in Turkey (51% pig), EB Apamee in Syria (37% pig).

Project Description:

This study adopts a novel approach to understanding social complexity in Bronze Age Greece. It uses the principles of Behavioral Ecology to study animals' interaction with their environment, especially in regard to human-animal interactions such as domestication[13] and animal husbandry. Using this framework, it explores factors that render an animal a good choice for exploitation - factors such as the social and economic effects that harnessing certain animals may have on a group of people, and the effects of domestication on local human ecology, economy, and social organization. For example, what factors affect the choice to manage cows versus pigs?

Preliminary study shows that pigs differ from domesticated bovids (cattle, sheep/goat) in a variety of distinct ways:

- Life history parameters
- Total production
- Secondary products
- Management economy
- Degree of domestication vs. management of wild population
- Social significance vis-à-vis their importance and use in ritual
- Differing role in local economy linked to settlement size

Additionally, faunal assemblages, pigs and small settlements, are all understudied in Greece, but all are vital in the reconstruction of Bronze Age society and culture.[14]

I. Taxonomic and Phylogenetic History of *Sus*

Since many studies focus on the domestication and origin of the pig, consideration of the taxonomy of the family, with a focus on the species *Sus scrofa,* is appropriate and necessary. Pigs belong to the order Artiodactyla and the family Suidae. Currently, the family Suidae, to which both wild hogs and domestic pigs belong, is divided into three subfamilies, five recent genera and eight to fourteen species (Nowak 1999). Wild hogs and

[13] The definition of the term "domestication" is a heavily debated issue. A wide variety of definitions exist depending upon whether researchers employ a biological or social framework. Given the interdisciplinary nature of anthropology, both frameworks need to be integrated to arrive at a working definition of domestication. Domestication is a biological phenomenon, but it is also a social and economic phenomenon as at its core is a human-animal interaction that transforms human-human relationships.

[14] The paucity of Greek faunal studies is clearly exemplified in a search of the Tozzer Index in which just 27 of the over 3000 faunal references pertain to Greece.

domestic pigs both belong to the subfamily Suinae, a group that includes *Sus, Potamochoerus,* and *Hylochoerus.* The remaining two subfamilies are the Phacochoerinae (*Phacochoerus)* and the Babyrousinae (*Babyrousa).* The genus under study in this paper is *Sus,* which will refer to both wild and domesticated forms. The rationale for grouping both domestic and wild pigs in one category is best justified with a look at their phylogenetic history.

As with most mammals, the taxonomic hierarchy of pigs has been the subject of debate. The family Suidae is commonly divided into five genera that correspond to geographic distributions worldwide. The genus *Sus* is comprised of eight species, with the species *Sus scrofa* having the widest geographic range (originally found in southern Scandinavia and Portugal to southeastern Siberia and Vietnam, North Africa, British Isles, and Asia). The remaining species of *Sus* are to be found in more limited and delineated areas. They include *S.salvanius* (SE Asian mainland), *S.bucculentus* (Vietnam, Laos), *S.verrucosus* (Java), *S.barbatus* (SE Asian islands), *S.philippensis* (Philippines), *S.cebifrons* (Central Philippines), and *S.celebensis* (SE Asian islands). Taxonomists differ in opinion on distinctions made in regard to domestic pigs, some choosing to place domestic pigs in their own species separate from *S.scrofa*, *S.domesticus*. Opposition to such differentiation highlights the fact that numerous populations of *S.scrofa* were probably involved in the origin of *S.domesticus* and that in some areas the wild and domestic forms can not be distinguished – a situation generally indicative of conspecificity (Nowak 1999). This examination will adopt the widely accepted position that both wild and domestic Eurasian pigs belong to the same species, *Sus scrofa.*

In addition to the preceding debate, some taxonomists have chosen to further subdivide the species *S.scrofa* into 4 subspecies (Genov 1999; Groves 1981). Animals are assigned to these subspecies according to morphological characteristics only, which happen to coincide with geographic distribution. These subspecies are *Sus scrofa scrofa* (Europe, North Africa, Near East), *Sus scrofa ussurieus* (southern Siberia to Japan), *Sus scrofa cristatus* (SE Asian mainland), and *Sus scrofa vittatus* (SE Asian islands). As this study is not a query into suid phylogenetics, it will suffice to mention that debate exists surrounding both the number of species and number of genera in the Suidae family, and that many of these debates are based upon morphological characteristics (to be addressed shortly) and geographic distribution.

II. Evolutionary History of the Wild Boar and the Domestic Pig

Recent advances in molecular genetics and research on mitochondrial DNA provide new avenues of inquiry into the origins of the human exploitation of pigs. Specifically, molecular based analyses might be able to provide researchers with the geographic origins of pig remains, thus shedding some light on questions surrounding the domestication of the species and early animal husbandry.

Phylogenies and evolutionary histories of the domestic pig in Europe point to an Asian wild boar ancestor. Yet, these systematic studies also mention the problems in simplifying the species history – a problem stemming from the long-term exploitation of pigs by humans. Evidence for *Sus* exploitation is plentiful from the beginning of the Neolithic, and so it is at least this point in history that evolutionary studies must begin. The central question regarding domestication is from where did the stock for domestic pigs come – and when.

A common explanation for the origin of the domestic pig is that domestic forms stem from the human exploitation of local, indigenous populations of wild boar. Wild boars were hunted, their young taken, and in turn used as founding populations for pig husbandry. Such practices appear to have occurred frequently throughout the species' domestic history, thereby rendering it difficult in some cases to ascertain wild from domesticated forms. Such practices render it necessary to look beyond morphological characteristics to molecular studies in order to address early domestication.

Study of sequenced mitochondrial DNA (mtDNA) and nuclear genes from wild and domestic Asian and European pigs suggests that the domestic pig originates from the Eurasian wild boar (*Sus scrofa).* It also indicates that domestication probably occurred independently from wild boar subspecies in Europe and Asia, with a time since divergence of the ancestral forms estimated at ~500,000 years (Guiffra et al. 2000). This date is well before early evidence of pig domestication at ~9,000 years. The results of a study conducted by Guiffra et al. (2000) indicate three distinct mtDNA clades in modern pigs, one Asian and two European. MtDNA sequences from some of the domestic pigs in the sample were closely related to European wild boar sequences, whereas others clustered with Asian wild boar sequences. This phenomenon "provides conclusive evidence for independent domestication of pigs in Europe and Asia" (Guiffra et al. 2000:1788). The study also found that there was a clear tendency for more pronounced allele frequency differences between populations from different continents than between wild and domestic pigs within continents, again supporting the notion that pigs were independently domesticated in Europe and Asia. Faunal evidence indicates that by 100,000 years ago, the wild boar was both abundant and widespread in Eurasia (Guiffra et al. 2000). An independent domestication of European and Asian wild boars followed by introgression provides a broad genetic basis for the domestic pig, and sheds light on the apparent confusion surrounding their place of domestication (Guiffra et al. 2000).

In addition, recent cytogenetic and biochemical investigations of wild boar chromosomes indicate at present two different species (or subspecies?) of wild boar, one with 2n=36 and the other with 2n=38. A few individuals were found to have 37 chromosomes, however modern domestic pigs tend to have 38 (Genov 1999). Interestingly, this molecular study appears to confirm what many earlier taxonomists and archaeologists hypothesized on the basis of morphological evidence – the domestication of the pig probably occurred repeatedly from local populations of wild hogs throughout varying geographic regions – and by different groups of people. Thus it does not seem likely that pig husbandry can be traced to one particular group of people or one specific geographic locale. Further examination of detailed aspects of pig domestication will illustrate the tenability of this conclusion.

III. Human Exploitation: Domestication and Management

Domestication:

Identifying domestication in the archaeological record is a subject that has garnered much attention over the years, with study focusing on cattle, sheep, goats, dogs, horses and pigs. Historically, these studies have relied primarily on examination of the morphological changes in a species diachronically, and only recently have molecular data from mitochondrial DNA been utilized.[15] Yet, the definition of domestication remains hotly debated. This debate has a profound affect on faunal analysis because it is only by starting with a definition that we can approach its recognition in faunal assemblages. Rather than provide a narrow definition of domestication, it is more beneficial to address the larger ideas

---

[15] Other combinations of factors also contribute to the zooarchaeological identification of domestication: changes in relative species abundances, changes in demographic parameters such as age distributions and sex ratios, changes in body size and shape, and the appearance of various pathologies (Berry 1969; Haber and Dayan 2004).

surrounding it. Therefore, we must begin by acknowledging that domestication is a process that includes a continuum of human and animal interaction (Haber and Dayan 2004).

Thus, we should not be looking for a specific date and place at which domestication of a species first occurred, but rather we should instead focus upon larger questions surrounding the processes of domestication that may be identifiable in the archaeological record. These processes are both biological and socio-cultural. Biologically, domestication may include artificial selection of certain traits by humans that can be maintained via effective reproductive isolation from the wild population (Bokonyi 1969; Haber and Dayan 2004). This human agency and control of the species in question is then a social and cultural process. However, a problem with the identification of domestication in archaeological contexts is that it is often made via observable morphological changes. In the absence of reproductive and genetic isolation from the wild forms, however, morphological changes due to domestication are difficult to see, and may have less of a chance of becoming established in a population. A lack of total genetic/reproductive isolation leads to the types of debate that surround the domestication of the pig – and may imply management of a wild population rather than complete reproductive control of a "domesticated" population.

The combination of genetic and behavioral changes affecting an animal due to domestication has a tremendous affect on the social and economic systems within a society. The challenge of addressing domestication from an anthropological approach is the necessity to integrate all of these affects. Biological definitions of domestication focus on genetic changes in animals as a result of selective and artificial breeding. Socio-cultural definitions may focus on the relationship between humans and animals and include practices such as selective culling of wild populations, capture of pregnant females in order to breed the young in captivity, and the raising of captured juveniles in a settlement for breeding stock that may be separated from the original wild population. Yet, neither the biological nor the social approach to domestication is sufficient (Russell 2002).

H. Hecker (1982) argued that the main hindrance in trying to define domestication lies in the way we have conceptualized the problem, essentially setting up a dichotomy between wild and domestic, and searching for evidence proving domestication's absence or presence. Conceptualizing the problem in this way overlooks three important points: "a.) animal domestication is first and foremost an adaptation and not a cultural imperative…b.) Local environmental factors must have played a critical role in whether a people developed or accepted the practice of domestication…c.) Domestication is a multidimensional process, the complexity of which cannot be reduced to a simple dichotomy" (H.Hecker 1982:218). Hecker suggests that the term "domestication" may actually interfere with the interpretation of the archaeological data. The reason for this interference is again, three-fold: "domestication" is an inadequate term to describe the complexity of the cultural behavior for which it was coined; the meaning has become imprecise because it has been defined differently by various authors;[16] there is a tendency to look for evidence of domestication and ignore other types of human-animal relationships (H. Hecker 1982). Anthropology is a multi-disciplinary field. The study of human animal interaction – the core of animal exploitation – must be studied from both a biological (genetics, behavioral ecology) and socio-cultural perspective. As a result, a succinct definition of domestication is not feasible, nor is it necessary. In this analysis, the term "domestication" is used in the classical sense, denoting intensive utilization of animals with selective breeding and reproductive control, along with isolation from wild forms – a definition that also encompasses management.

One possible source of confusion is that we are unclear on the sources of variation in key behavioral and physical traits. For instance, assume a large part of "domestication" is "taming" of an animal. Is taming achieved by selection of specific alleles that contribute to an animal's temperament, or is it achieved by altering the way an animal is raised? Is "bad temperament" a matter of 'nature" (genes) or "nurture" (environment)? As evolutionary psychologists and behavioral biologists study this question in our own species, it has become increasingly apparent that the very premise of the question … the so-called "nature-nurture dichotomy" is itself a fallacy. If that is the state of understanding the biological basis of personality in humans, then we must admit that our understanding of domestic animals is truly at an impasse.

Management:

Heavy exploitation of animal resources that eventually lead to complete human control over population reproduction can certainly be called "domestication", but applying this term indiscriminately to fauna overlooks the idea that human exploitation of animal resources is a continuum of practices – not a presence/absence dichotomy. Traditionally, faunal analyses have focused on the identification of either wild or domestic forms of a

[16] Bokonyi (1969:219) offers a classic definition of domestication as "the capture and taming by man of animals of a species with particular behavioral characteristics, their removal from their natural living areas and breeding community, and their maintenance under controlled breeding conditions for profit". Webster's collegiate dictionary, 10th edition defines domestication in broader terms "to adapt (an animal or plant) to life in intimate association with and to the advantage of humans".

species in assemblages, but this dichotomy may be artificial, and unsuitable to every culture and site. It may be beneficial to forgo this dichotomy, and instead view human-animal interactions as a continuum of practices. By abandoning the assignation of wild or domestic to fauna, we can ask different questions of bone assemblages. The need to consider alternative frameworks can best be highlighted by examining how "domestication" is traditionally identified, and the problems inherent in this identification in relationship to pigs.

Studies of pig domestication reveal that heavy stress has been placed on morphological changes, and this contributes to the murkiness surrounding understanding of their human exploitation (Bokonyi 1969; Genov 1999). Pigs are ubiquitous, and indigenous to most parts of the Mediterranean. There is great advantage to allowing wild populations to graze freely outside settlements – a practice that would maintain gene flow between wild and more managed populations. With the maintenance of gene flow, morphological changes would be difficult to see, and may even fail to occur. Thus, what may have occurred was management of wild populations of pigs, rather than full-fledged "domestication". Focus on morphology alone overlooks the processual aspect of domestication and instead renders it a dichotomy – a dichotomy that is not present in all faunal assemblages.

Management is an issue especially relevant to this research, as pigs are better suited to such a system than the other major domesticates. Since pigs are generalist omnivores who thrive on being allowed to roam freely in areas that provide ample food resources, they need not have been a fully domesticated resource. Humans may instead have managed wild populations by occasionally harnessing/capturing wild females when needed, maintaining ample gene flow between wild and "domestic" individuals. This theory might explain the lack of clear morphological changes commonly attributed to domestication in pigs.

The preceding ideas provide some insight into the issues encountered in the identification of domestication in general, and pigs in particular. The application of generalities usually applied to species domestication, such as reproductive isolation, do not work so well when applied to pigs. With respect to genetic and reproductive isolation, early domesticated pigs frequently experienced what can be labeled "short-term" reproductive isolation, as there appears to be evidence for somewhat frequent genetic exchange between wild and domestic forms. However, since changes in morphological characteristics are the traditional way in which divisions have been drawn between wild and domestic pigs, it is to a discussion of these traits that this chapter now turns.

Morphological Approaches to Domestication:

Despite the theoretical issues surrounding the identification of domestication and management in pigs, it remains beneficial to examine some of the morphological characteristics that have traditionally been employed in the discernment of wild and domestic populations. The following discussion delineates changes that have traditionally been regarded as a response to, and a result of, domestication.

A general trend in domesticated animals is a decrease in body size from their wild forms diachronically. This occurs in cattle, sheep, goats, and even in attempts to domesticate deer. Domesticated pigs do not always adhere to this rule. In fact, faunal remains in many archaeological contexts indicate that domesticated pigs tend to become larger in prehistoric periods, smaller in Antiquity, and then larger again with increased animal husbandry (Haber and Dayan 2004). We know from Greek and Roman texts that the practice of fattening pigs was common during Antiquity, but we cannot be certain when this practice first began. This phenomenon is most problematic when trying to identify and differentiate wild from domestic pigs in early archaeological contexts. Both of these factors render generalizations surrounding early pig domestication difficult.

Other size indicators of domestication may exist, however, and research has identified those morphological characteristics that may at times be useful in distinguishing between wild boar and domesticated pig in archaeological faunal assemblages. Several individuals have examined this issue from various perspectives, with the most frequently studied element being the cranium, and specifically the relative size and shape of the lachrymal bone (Kelm 1938; Major 1883; Nathusius 1864). In general, the shortening of the lachrymal bone, the overall shortening of the snout and a reduction in tooth size with crowding (Bokonyi 1974) have all been utilized as indicators of domestication.

Due to the confusion surrounding discernment of wild from domestic on size alone, Payne and Bull (1988) analyzed this dichotomy in pigs, and set out those measurements they found most reliable in addressing this issue. They postulated that variation within an animal population could be thought of as made up of three components:

1. Age related change: changes in size/shape during the life of an animal; measured by differences in mean measurements between samples of animals of different ages (most increase with age, except tooth crown height); post-cranial measurements are larger in older individuals, especially forelimbs (up to 14 percent)

2. Sexual dimorphism: measured by differences in mean size between males/females of same age (castrates may differ); greatest differences in forelimb measurements (8-12 percent)
3. Residual individual variation: remaining variation shown by a sample of animals of same age/sex from a single population (includes genetic and environmental components and observer error); computed by using Pearson's Coefficient of Variation (COV) which expresses the Standard Deviation (SD) as a percent of the mean → v=sd x 100/mean; low variability with COV 1-5 percent; In general, in a modern sample the COV ranges from 3-9 percent, tending to be higher for post-cranial measurements than for teeth, and especially low for tooth width measurements.

	Age-related change	***Sexual Dimorphism***	***Residual individual Variability***
Forelimb measurements	Increase	High	Low/moderate
Hindlimb measurements	Some increase	Moderate	Low/moderate
Tooth widths	Relatively stable	Nil/low	Low/moderate
Tooth lengths	Decrease	Nil/low	Low/moderate

Table 5.1 **Variation in pig measurements (after Payne & Bull 1988)**

Payne and Bull (1988) suggest that in order to separate larger wild from smaller domestic animals, or to investigate size change between different periods, suitable measurements should ideally have low sexual dimorphism, low age-related change, and low residual individual variability (*Table 5.1*). The most common measurements used in published studies for separating wild and domestic pigs is tooth length, especially of the third molar, and various post-cranial measurements. Tooth width, however, which shows less age-related change than length and less sexual dimorphism than postcranial measurements, was found to be most useful in determination of wild and domestic forms. Third molar length is less affected by wear than first and second, but third molars are scarce in archaeological collections, and may have relatively high residual variability.

Since post-cranial measurements show considerable sexual dimorphism and age-related change, and thus high total variability, they are less effective at distinguishing between wild and domestic fauna. If we must use postcranial measurements, then we should look for measurements with high sexual dimorphism and low age-related change, such as the humerus HTC (diameter of trochlear constriction). Pelvic measurements may also be useful. In general, however, the data suggests that tooth width and hindlimb measurements are likely to show lower overall variation and thus discriminate better between larger wild and smaller domestic pigs than tooth lengths and forelimb measurements.

> Even though there may be significant differences between measurements for separate groups, such as wild and domestic pigs, or males and females, it may in practice not be possible to separate them in mixed collections. When the means of the two groups are separated by less than two standard deviations, the two distributions will merge to form a single broader mode. Bimodality only starts to be apparent when means are separated by four or five standard deviations (Payne and Bull, 1988:32)

Genov (1999) reached similar conclusions to Payne and Bull (1988), utilizing classical biometrical analyses, with the aim of clarifying, which, if any, morphological characteristics might be useful in species identification and questions of domestication. He concluded that a distinction between wild and domestic forms cannot be drawn so facilely, due to the probable frequent genetic mixing of wild boar and domesticated pig. Since the domestication of the pig probably occurred independently in several parts of the world at various times from indigenous wild boar, morphological generalizations are tenuous and genetic analyses complicated. We can thus turn to evidence from written texts as another avenue of inquiry.

Textual References in Antiquity:

In addition to molecular and morphological studies, early textual references also provide information on ancient pig utilization. The earliest written mention of the exploitation of pigs comes from Ancient Greece, specifically from the Linear B tablets found at Pylos from the Later Bronze Age. While mention of pigs is scarce, evidence does suggest that they were a part of the palatial economy. They are listed as having been sent to the capital, however in relatively small numbers. The tablets divide pigs, like other mammals, into male and female, and indicate that certain of these had been fattened, whether for meat or sacrifice is unclear (Mancz 1989). Pig skins are also accorded mention in the tablets in relation to the leather-working industry. Of particular interest in the Linear B tablets is the small role accorded to pigs in the palatial economy. The tablets themselves may reflect a dichotomy between the actual relative frequencies of pig remains at archaeological sites near the palace, in particular Nichoria, and their actual significance to the larger palatial economy. The heavy local exploitation of pigs might suggest that they were important to smaller scale societies as food staples, but unimportant as a resource for the larger wealth of the palatial economy (Mancz 1989). A heavy dependence on pigs may be indicative of a less wealthy settlement, an idea which will be addressed in chapters 7 and 8. It is also worthy of mention that in addition to the tablets from Pylos, Homer's *Iliad* and *Odyssey* may also be used to provide clues to the later Bronze Age exploitation of pigs. Both works contain numerous references to pigs and their scrutiny might shed light on early cultural aspects of pig husbandry. During even later periods of ancient Greece, pigs are frequently found as sacrificial offerings in ritual contexts, and in particular they are heavily associated with the cult of Demeter (goddess of corn, akin to Ceres), nearly always having a presence at any shrine or temple dedicated to her.

While relatively little has been written on the role of the pig in Greece, there exists a fair amount of research on its role in Roman Italy. It is in this area that we find a focus that is not entirely zooarchaeological, but of a more literary and artistic nature. Roman art abounds with depictions of both what have been taken to be wild boar and domestic pigs, and Latin classical sources (Apicius, Virgil, Pliny, Varro, Calumella, and Palladius) provide a wealth of advice on animal husbandry, sacrificial proscriptions, and even recipes for the more discerning. Of note is Mackinnon's (2001) study on the marriage between zooarchaeology, literature and art in respect to differentiating pig breeds in Roman Italy. Utilizing the above approaches, Mackinnon identifies the presence of at least two different breeds of pig: a large, fat, short-legged variety, and a small, bristled, long-legged variety. While limited in size and scope, Mackinnon's study can be a valuable reference against which to compare hypotheses on the role of the pig in other areas of the Mediterranean diachronically. It can be argued that even given the nature and bias of classical authors, their insights may be a more valuable yardstick in assessing the economic and social role of pigs than a more modern comparison. Thus, we should consider what the classical sources have to say in order to provide added insight into early animal husbandry techniques, which, when viewed in conjunction with the faunal record, may be an aid in the study of the socio-economic importance of pigs in antiquity.

Given many of the issues associated with the determination of the differences between wild and domesticated forms, it is of no surprise that little has been written about them beyond note of their presence in faunal assemblages. However, what is important is not how many breeds or species of pig there are, or when, where and by whom pigs were domesticated, but rather the characteristics that were selected by humans. An examination of desired traits can then be employed to shed light on aspects of early animal husbandry – ultimately enabling us to better understand the social and economic climates of the peoples involved. Perhaps what is of greater importance is that the presence of the pig in early animal husbandry illustrates that what spread was the idea of domestication. Thus, if it is the "idea" of domesticating animals that spreads then this supposition may lend support to the conclusion that domestication occurred in differing geographic areas independently. Instead of the introduction of new domesticates via human translocation or even trade, perhaps the simultaneous independent and local domestication of the pig should be used as an indicator of the spread of ideas. In the case of domestication, this spread of ideas had a profound and altering impact on the peoples whose ears it reached. Thus, maybe we should be examining the role of the pig in archaeological contexts as a social and economic barometer and focus our attention on what such studies can illuminate.

Archaeological Evidence of Domestication/Management:

Before examining why pigs are such a wise choice for human exploitation, a brief glance at some of the archaeological evidence used to approach questions of domestication can add to this discussion. Archaeological evidence from Greece and the Near East, which is corroborated with molecular data, indicates that heavy exploitation of pigs could have occurred as early as 9,000 years ago, and was certainly in place by the time of early sedentary societies (Bokonyi 1974; Tikhonov et al. 2004). Faunal evidence for pigs in early archaeological contexts is quite frequently limited to mention of their presence. It is only when we are fortunate enough to examine something like the change in kill-off patterns at a site diachronically that we can begin to truly test assumptions of "domestication". Such a study was conducted at the prehistoric site of Hagoshrim in northern Israel, where examination of pig remains indicates a diachronic change in culling patterns, with a clear pattern of a younger culling age during later time periods. This implies a larger population of younger pigs, suggesting domesticated stock (Haber and Dayan 2004). Hagoshrim demonstrates the potential archaeological contexts may hold in regard to questions of domestication, and yet it is one of the few Mediterranean sites thus far to be examined within this framework.

In addition to Hagoshrim, faunal evidence for pig exploitation is ubiquitous in many ancient settlement contexts in the Mediterranean, fuelling the debate concerning the time and place of the earliest pig domestication. Extensively excavated settlements, such as Jericho, Argissa-Magula, Paradeisos, Lerna, Mycenae, Knossos, and Pylos, to mention but a few, have evidence of pig domestication well in place by the Neolithic. This still leaves unanswered questions of timing and locale. Perhaps it is because so much evidence exists for their early utilization both geographically and diachronically that pinning down a place in time and space is so difficult. Fortunately, a study with a focus on the social and economic importance of pigs does not necessitate finding an answer to this chronological and geographical quest. While it would be wonderful to be able to assign a date and a place to their domestication, in the larger analytical framework, it is not necessary. Scholars have focused on those questions for over a century, and they have not served to further our understanding of the larger socio-economic influence of pig husbandry on societies.

IV. Why Pigs?

In order to fully address the importance of pigs in antiquity, we must examine the specific life history variables that render them such a valuable commodity to animal husbandry. As an in-depth discussion of behavioral ecology is presented in Chapter 9, discussion here will be brief, providing the bare background information necessary to support the ensuing discussion of hypotheses concerning pig exploitation.

A cursory glance at the very small number of species that have been chosen for domestication by humans consistently turns up similar criteria. Sheep, goat and cattle serve a dual purpose in that they are traditionally exploited for both their primary and secondary products. Extensive research has been conducted on these species from which we are now able to ascertain the probable function of the occurrence of these animals in archaeological contexts. In a broad sense, pigs are an ideal species for domestication. They are a generalist species with high reproductive potential, broad food selection and habitat generalization. From various aspects, pigs resemble dogs and humans more so than cattle or sheep and goat. As omnivores, they can be fed a wide variety of food, including bone, so they can be fed with leftovers from human use, and they can be used to help eliminate rotten food from settlements (Haber and Dayan 2004). Additionally, their fat content is higher than other domesticates. This coupled with their high reproductive rate, renders them a natural choice for domestication. Yet with respect to human exploitation, one of the traits that renders pigs unique from other commonly recovered domesticates is that they primarily serve one function – they are used for their meat. Although their hide and tusks are also utilized, they have no other secondary products to offer, like the milk produced by the three other commonly domesticated species – sheep, goat and cattle.

Perhaps the greatest negative aspect of pig domestication is that in times of scarcity, they may be in direct competition with humans for food resources, frequently having been known to destroy crops. Competition with humans appears to have been, and still is, greater in arid environments, and perhaps this is one of the factors explaining their more limited exploitation in parts of the Near East and Northern Africa, as compared to their high favorability and frequency in Europe. During Egypt's Middle Kingdom, pigs were thought to represent everything that is evil and were partially banned, a practice that may well have stemmed from their destruction of crops vital to humans (Haber and Dayan 2004; Redding 1991). Given each of these factors, we should then inquire why the frequency of pig remains is just as high, if not higher, at many archaeological sites throughout the Mediterranean and Europe as these other, seemingly more beneficial animals.

When examining the relative benefit of domesticating a certain species we can examine what the species in question can offer via subsistence, and conversely, what it will necessitate to care for and sustain until the point of slaughter. One factor that sets pigs apart from other domesticated animals is their high number of offspring and short time to sexual maturity. A sow averages eight piglets per litter, with some capable of bearing up to 12. Compare this to the single litters of sheep, goat and cattle, and we begin to understand why pigs appear so frequently in the faunal assemblages of archaeological sites. In addition to bearing a large number of young, a sow can reach sexual maturity during the period from 7 to 22 months, while males reach sexual maturity by their tenth month (Dobney 2000; Tikhonov et al. 2004). Although wild boars usually bear one litter per year, pigs are capable of bearing two, as modern sows are in heat for a period every three to four weeks. Latin texts also speak of pigs being capable of producing two litters per year (Varro *Rust.* 2.4.7: Palladius 3.153; Columella 7.9.4), although advising against such practices (Lauwerier 1983; Mackinnon 2001).

Recent molecular studies can also provide explanation as to the rationale for pig husbandry. Tikhonov et al. (2004) compared the genotypic structures of most globally domesticated populations of breeds and found that heterozygosity in genotypes actually increased the number of piglets born per litter. Their research found that in addition to increasing prolificacy (including multiple pregnancies), pigs with an Asian fund of alleles in the genome and an increased heterozygosity in their genotypes also had greater pre- and postnatal viability (providing minimum mortality) and acceleration of somatic growth and maturation (Tikhonov et al. 2004). Although molecular study of this nature has not yet been performed on pig remains from archaeological contexts, it is possible that early practitioners of animal husbandry were aware of the ability to positively affect the litter size of managed wild boar, thus rendering them highly valuable as both a food and economic commodity.

Ecology also contributes to our understanding of the reasons behind the widespread exploitation of pigs. A cursory glance at the global distribution of indigenous wild boar, from Western Europe to the Far East (including many islands), illustrates their suitability to most temperate regions of the world. Large populations of feral pigs can also be found in areas such as Australia, the Americas and islands such as New Zealand, and this distribution coupled with the expansion of the range of the wild boar, illustrates their ability to thrive even in areas heavily influenced by human activity (Schley and Roper 2003). Their habitat preference, however, is for forested areas that provide cover, shade and a source of food and water. Study of the density of wild pig populations suggests that access to riverine woodlands is the determining factor in group size and habitat selection (Choquenot and Ruscoe 2003; Virgos 2002). Specifically, research suggests that proximity to riverine woodlands constrains foraging, since riverine woodlands contain better quality pasture than the surrounding shrublands. However, an attendant factor to this finding is the role of thermoregulation. Lacking sweat glands, pigs die if exposed to full sun when ambient temperatures exceed 23° C, or partial sun when ambient temperatures exceed 35° C. (Choquenot and Ruscoe 2003). Thus, there is a vital need for thermoregulation in warmer climates and their indigenous habitat appears to be governed by this need. Riverine woodlands fill this need, and they occur in most geographic locales. Preference for more sheltered habitats may impart other benefits as well, such as refuge from predation or hunting and ready access to water. It is also noteworthy that the Mediterranean climate through the early Bronze Age was more temperate than in the present day (Dobson 1998), and this may have also played a role in the development of pig husbandry. This is an important factor that merits a few words.

Climate Change:

An important component of understanding the diachronic change in faunal remains of any species and at any site is a reconstruction of the paleo-environment and climate change – both factors that can shed light on a shift in the exploitation of a specific fauna. An example of this paradigm is the environmental changes that occurred in the Mediterranean region from the Neolithic through the Bronze Age. Avian faunal evidence, as well as geological evidence, suggests that the period before and during the early part of the Neolithic was more arid than today, with less rainfall. During the end of the early Neolithic, this same evidence indicates a shift toward a wetter, warmer, and more humid climate, extending into the late Bronze Age (Gejvall 1969). This type of environment appears to correlate with the rise – and fall – in pig exploitation. The pattern of pig exploitation in the Neolithic seems to mirror the changes in climate. The early Neolithic reflects a pattern of less dependence on pigs, and is also a more arid period climatically, while the later Neolithic through the Middle Bronze Age is one of rising, heavy pig exploitation. This coincides with a wetter, warmer climate, one that is an ideal natural habitat for swine. During the end of the Bronze Age, the climate appears to become dryer, colder and more arid, and this appears to coincide with a decrease in the frequency of pig remains archaeologically (Gejvall 1969; Mancz 1989). Mention of these data patterns here are cursory, but the correlation between climate change and the rise and fall of pig exploitation certainly warrants further exploration.

Dietary considerations:

The omnivorous diet of pigs is also a contributing factor in their suitability to many areas – and arguably must have factored importantly in considerations of early domestication. Studies of modern wild boar indicate that their diet consists of a variety of vegetable foods, such as mast, roots, green plant matter, and agricultural crops. In addition seeds, nuts, fruit, vertebrates and invertebrates comprise their diet (Shley and Roper 2003). The wide variability in the diet of wild boar is generally interpreted to mean that they are opportunistic omnivores, whose diet in any particular instance is strongly influenced by availability. This adaptability is one of the driving factors in their widespread use and domestication diachronically.

Classical written texts also contain instructions for the diet of domestic pigs, which include the feeding of scraps and litter (Plin. *NH* 8.77.206). Taken in concert, the ecology of *Sus scrofa* enables us to see just how rationale a choice domesticating the pig was. Pigs are animals whose wild form is ubiquitous, present in areas of human impact, have large litters, provide a good amount of fatty meat, reach sexual maturity quickly, and have the ability to sustain themselves with a wide variety of foods. We must now ask, however, what additional factors surround their suitability for domestication.

Unlike cattle, and arguably sheep and goats, pigs do not require a large area in which to be kept. There is also ample evidence suggesting that, unlike other domesticated animals, pigs experienced an increase in size due to domestication, apparent in archaeological contexts beginning in the Neolithic and continuing to the late Bronze Age (Bokonyi 1974). They are also capable of subsisting on the leftover food produced by human activity, and so are not dependent upon any particular environmental food resource. Perhaps their largest requirement is shelter from extremes of hot and cold temperatures. Roman period archaeological evidence indicates the existence of sties for the keeping of pigs (Mackinnon 2001), and ancient literary sources confirm their function (e.g. Pliny *Naturalis Historia*). From an economic and practical perspective, a small sty can be easier to attain than a large pasture for many early cultures. Most importantly, however, the keeping and domesticating of pigs is a reflection of a sedentary society. Pigs are animals well suited to small-scale agricultural communities, and so perhaps the question that should be addressed is whether the domestication and utilization of the wild boar played a role in the beginnings of sedentism. Much focus has been given to the role of agriculture in the formation of larger scale, sedentary societies, but the domestication of a locally available, practical source of food must also have contributed.

V. Exploitation of *Sus scrofa* in the Ancient Mediterranean

At this point, several factors regarding the suitability of pigs as domesticates are apparent. They are indigenous to many areas in which early human settlements first appear, they are naturally suited to most temperate, wooded environments, they are generalist omnivores with an ability to be sustained on a wide variety of foods, they are naturally capable of having up to two farrowings per year and bearing large litters, and they do not require a large amount of space. The combination of these factors renders the pig a natural choice for exploitation. In order to better understand the role of pigs as important aspects of early social and economic sedentary societies, we must draw conclusions about their socio-economic utility based upon evidence drawn from archaeological contexts. One approach to conducting such an inquiry is to examine a few sites with heavy pig exploitation in order to see if it is possible to shed light on their larger role in early society.

Here I will present brief synopses of early pig exploitation in the Mediterranean in order to further highlight the need for new inquiries into the important role of the pig in early social and economic contexts. In the southern Levant, high frequencies of pigs first appear by the end of the Pre-Pottery Neolithic (PPN) roughly 8000 BP, almost always being identified as wild (Haber and Dayan 2004). Domesticated pigs, an attribute based upon morphological size reduction, appear during Wadi-Raba times, nearly 2000 years later. Similar conclusions have been drawn at many other sites in Turkey and the Near East, such as Hagoshrim, Jarmo, Gurcutepe, Hayaz Tepe, Tel-Halula and Cayonu Ervynck (Haber & Dayan 2004; Rolett & Chiu 1994). In each of these cases, conclusions of domestication were drawn from both morphology and kill-off patterns.

In Greece, there is an increase in pig remains also beginning in the Neolithic, from Nea Nikomedeia and Argissa Magula in particular. However at the majority of Greek sites, strong attempts are not made to distinguish between wild and domestic forms of pigs.[17] This is partially due to the practice of allowing the pigs to roam in the woodlands, assuming genetic mixing with wild forms (Larje 1987). Additionally, it has been proposed that pig stock was replenished periodically from local wild forms – a practice that renders distinctions difficult, but one that does confirm local and independent domestication (Larje 1987). At the Late Neolithic Greek site of Paradeisos, pig remains are the second most frequently occurring species comprising around 23 percent of the livestock, preceded only by ovicaprines and followed by cattle. Of particular note, however, is the pattern of increased frequency of pig remains from the Early Neolithic to the Late and End Neolithic. This pattern mirrors frequencies of pig remains from additional sites in and around Greece diachronically, such as at Chevdar, Nea Nikomedeia, Lerna (Gejvall 1969), Argissa Magula, Sitagroi (Bokonyi 1986), Agia Sofia-Magula, Kazanluk, and Magula Pevkakia (Larje 1987). Similar patterns are also seen in the Near East. Given the increased frequency of pig remains from the Early Neolithic through the end of the Bronze Age (at sites in which we see continuous exploitation), we need to understand what factors might have accounted for this shift. The following pages will examine models that seek to explain and account for variation in the intensity of pig utilization.

VI. The Model

This study suggests the following model: Pig utilization in early Bronze Age Greece varied in relation to important aspects of social organization, economics, and agricultural practices. Faunal studies of pig use can be used to address inter-site economic relationships, as well as intra-site subsistence economy. In a sense, pigs may serve as a "yardstick" for

[17] Gejvall's (1969) study of the fauna from Lerna is an exception to this trend.

important socio-economic practices or changes in these practices. One difficulty, however, is the problem of equifinality. On a given site the increase/decrease in percentage of pig may be caused by many things. It is therefore important to parse out and carefully examine the various likely or plausible relationships between pig utilization and socio-economic or technological change. In order to address this problem, a series of specific hypotheses linking pigs to socioeconomic organization in EH Greece are here assembled from the archaeological literature. Wherever possible these are phrased in terms of behavioral biology because strong theory exists to help define some of these relationships. This hypotheses and its ensuing models will be tested in chapter 8 through application to several Early Helladic sites in the Peloponnese (Lerna, Helike, Tsoungiza and Tiryns), and then comparing these with Late Helladic sites for which we have better knowledge of social organization. The following hypotheses will address Early Helladic social organization as it related to mobility/sedentism, agricultural intensity and intra- and inter-site economic relationships (degree of centralization of a society), and class (social stratification).

Hypotheses:

Several hypotheses have been used to explain the intensity of pig exploitation in archaeological contexts and ancient societies. Here I present each hypothesis and illustrate these various intrinsic interwoven aspects. For each hypothesis, there is a culture-historical, economic, agro-technical, and behavioral biological aspect. These hypotheses are reassessed in chapter 8 and applied to the specific Early Helladic faunal assemblages under analysis.

- *Ecological hypothesis*: Successful husbandry of pigs depends on specific geographic and climatic factors, such as the amount of rainfall in an area. Specifically, Grigson (1987, 1995) has suggested that as rainfall increases, pig use increases, with a lower threshold of 300-350 mm annual rainfall, and a presumed leveling off at some higher number. If annual rainfall levels are less substantial infrastructural investments would need to be made.

Zeder's work at Tell Halif (1996) addressed and tested the ecological, behavioral and political aspects of the ecological hypothesis.

> Ecological theories for the paucity of pork in the Near Eastern diet have concentrated on physiological characteristics that might affect the fitness of pigs raised in arid environments. Some maintain that the relatively high water requirements of pigs make them ill suited to semi-desertic areas (Grigson 1987; Horwitz and Tchernov 1989). Harris argues that because pigs lack sweat glands which allow other Near Eastern domesticates to adjust to high temperatures, the labor investment required to keep pigs cool in arid environments outweighs their

benefits (Harris 1985:74-75). Furthermore, without the ruminant's multi-chambered, bacterially-fueled, grass-processing stomach, pigs are unable effectively to process plant foods high in cellulose. Since the primary fodder plants in the Near East are cereals, domestic pigs would seem to be at a definite disadvantage to cud-chewing sheep, goat and cattle (Zeder 1996:298).

- *Mobility (transhumance) hypothesis*: The more mobile a society, the less likely it is to utilize pigs as a major resource since pigs are difficult to herd (Flannery 1983; Harris 1985; Zeuner 1963). Pigs are ill-suited to nomadic pastoralism, and thus increased frequency of pigs at a site may indicate increased or complete sedentism. Even a small number of pigs must indicate either some sedentism or links to sedentary societies.

"Behavioral theories have concentrated on the contrary nature of the basic swine disposition. Because pigs are temperamentally ill-suited to herding, Flannery (1983), Zeuner (1963:260)…have postulated that pigs are not compatible with a more nomadic way of life" (Zeder 1996:298).

This chapter has illustrated how both biological and ecological variables can explain the suitability of pigs for human exploitation, but there are also other factors that must be considered. Unlike cattle and arguably sheep and goats, pigs do not require a large area in which to be kept. From an economic and practical perspective, a small sty can be easier to attain than a large pasture for many early cultures. Most importantly, however, the keeping and intensive management of pigs is a reflection of a sedentary society. Pigs are animals well suited to small-scale agricultural communities, and so perhaps the question that should be addressed is whether the domestication and utilization of the wild boar played a role in the beginnings of sedentism. Much focus has been given to the role of agriculture in the formation of larger scale, sedentary societies, but the domestication of a locally available, practical source of food must also have contributed.

- *Agricultural intensification hypothesis*: The intensification of grain production is accompanied by a shift from pig husbandry to one focused on cattle and goats (Redding 1991). Pigs are destructive to crops and must be kept out of agricultural areas. In addition, there is a competitive relationship between pigs and humans for resources. Thus, a site with high intensity agriculture would benefit by utilizing an animal that is more compatible with such a system, such as goats and sheep. In addition, an increase in the frequency of cattle should accompany this decrease in pigs as cattle can be used as work animals for plowing necessary in agricultural intensification. However, even in an intensive agricultural system, pigs could be utilized in a major way if further specialization/differentiation arises.

Because pigs consume foods capable of supporting human populations, some have postulated that in arid areas, pigs may become a serious competitor for scarce food resources (Harris 1985). Redding (1991) has argued that although pigs are not able to process cereals as efficiently as bovids, they do raid gardens and venture into fields, consuming ripening heads of wheat and barley, trampling and uprooting plants. The invasion of agricultural fields by unpenned pigs can cause extensive damage, placing pigs in direct competition with humans. As a result, Redding predicts that dependence on pigs in the Near East should decrease with the increased intensification of agricultural production (Zeder 1996:298).

- *Centralized society/Urban economic hypothesis (Political economy)*: The frequency of pigs should be inversely correlated to the degree of centralization of a society (Diener and Robkin 1978). Variance among sites in pig use should increase. Pig utilization is a good strategy for small, independent settlements. Pigs yield a high amount of meat and can be easily fenced out of small, garden cultivation areas, enabling a community to maintain independence from larger, 'administrative' centers or centers of authority. Pig husbandry may be a useful rural subsistence strategy that permits satellite communities to emphasize their domestic, non-market based modes of production and, in so doing, maintain a degree of autonomy from those centers in the political economy which seek to control them (Deiner and Robkin 1978). This hypothesis is related to the agricultural hypothesis and also touches on the idea that pigs may be an initial subsistence strategy and fall in disuse as a site matures into a larger center.

Diener and Robkin (1978) postulate that those seeking to control the flow of commodities to dependent consumers would not welcome the autonomy small-scale pig raising provides independent households. As a result, pig rearing would be actively discouraged in centrally coordinated state-level economies. Recently Hesse and Wapnish (Hesse and Wapnish 1987; Hesse 1988) applied this theory to an examination of Middle to Late Bronze Age Levantine subsistence economies, arguing that pig utilization in the region is inversely related to the growth of a large centralized olive processing industry and to the degree of political and economic domination by external powers (Zeder 1996:299).

- *Class hypothesis*: The production/consumption of pork is associated with lower or working class status (R.T. Hecker 1982; Panagiotakopulu 1999). Pigs reproduce quickly, are omnivorous, and would be a cheap and relatively efficient way to feed workers. Additionally, small household could afford to keep several pigs as food sources.

Each of the above hypotheses has been, and can be, used to both predict and explain the frequency of pigs in ancient settlements. Each hypothesis has both an intra-site component and inter-site component, so they not only allow us to examine social organization at one site, but across sites. Chapter 8 applies each of these hypotheses to the Early Helladic faunal assemblages from Helike, Tsoungiza, Lerna and Tiryns in order to test the model that pigs can act as a litmus test of economic organization, and thus a proxy indicator for social complexity.

VII. Predictions, Hypotheses and Data Classes

Faunal evidence suggests that pigs played a significant role in the local economies of Bronze Age settlements in Greece. This study addresses the following:

- During this period, and perhaps for much of prehistory, pigs are not fully domesticated, nor wild, but are instead managed over time in a state of flux between these two extremes. This strategy would work effectively in smaller settlements. However, in larger settlements, stricter boundaries are drawn around private property and private ownership of land, promoting a strategy of ownership over animals and property. In smaller villages with community property, it may not be as necessary to privately "own" pigs. These ideas and hypotheses can be tested via collection and analysis of the following data classes:[18]
    - compare relative numbers of pigs recovered in small and large settlements via quantitative analyses: NISP, MNE (MNI, MAU)
    - examine observable changes in skeletal element frequencies and body part distribution
    - examine age profiles, including culling patterns
- A significant difference exists in the intensity of pig exploitation between small settlements and larger administrative sites. Pigs are heavily exploited in areas in which garden cultivation forms the main subsistence strategy, and are less heavily exploited in areas with intensive grain/cereal agriculture. Thus, in small settlements pigs may play a larger role in subsistence than in larger settlements. Accounting for

[18] See Ioannidou 2003; May et al. 1996; Payne and Bull 1988; Prummel 1987, 1988, 1989; Weinstock 1993

this difference may be that pigs are known to compete directly with humans for food resources, and in areas with large scale agriculture, it is difficult to prevent pigs from destroying crops. In areas with smaller, garden cultivation, pigs can be fenced out of cultivated areas, and allowed to roam freely outside the settlement. These ideas and hypotheses can be tested via collection and analysis of the following data classes:

- compare quantitative data between large and small settlements (NISP, MNE, etc)
- compare qualitative data between large and small settlements: presence/absence of bone surface modification (pathology, trauma, butchery marks, burning)
- compare skeletal element frequencies between sites for evidence of utilization

➢ The exploitation of pigs is inversely correlated with the socio-economic complexity of a society. The more centralized the society, the less reliance on intensive pig utilization. Aspects of social complexity typically include the heavy exploitation of secondary animal products, such as milk and wool, intensive agriculture, craft specialization and a dense population.[19] This inverse correlation may be tested by examining the potential reliance on secondary products as a proxy indicator for trade and centralized society through the age structures of sheep remains. A diachronic comparative analysis can be done with Late Bronze Age Mycenaean sites known to have exploited the secondary products of sheep and goats, examining their relative numbers of both these species and pigs, and then comparing them to early Bronze Age sites, in which the extent of societal centralization is unknown.

VIII. Conclusion

This summary of some of the factors governing the exploitation of pigs in early social systems illustrates the dire need for a more integrative approach to the zooarchaeological record, especially in Greece and other areas of the Mediterranean.

Current and extant faunal analyses in Greece fail to examine faunal evidence against the larger socio-economic context, of which they are surely a part. Pigs played a significant role in the subsistence and economies of early sedentary societies, and were among the

---

[19] Evidence from the 4th millennium early Bronze Age assemblage at Zeytinli Bahçe Höyük (Urfa, Turkey) suggests an increase in the utilization of pigs occurs in less centralized cultures (Siracusano 2004).

earliest managed and domesticated animals. Their numbers are consistently significant both diachronically and geographically, and they are frequently represented artistically, written about, and are the subject of sacrificial offerings. Given their abundant appearance in so very many aspects of daily life, it is difficult to believe that few attempts have been made to understand them within a socio-economic framework. Rather than maintaining a focus on questions surrounding when and where the first pigs were domesticated, we should turn to what their presence signifies for early societies. What role do they play in the greater economic subsistence of the settlement? What is their larger role in a subsistence economy? Are they indicators of particular types of societies (e.g. small scale, large scale, urban, rural)? How were they viewed? Each of these questions, and more, need to be addressed before we can truly understand the role of animal husbandry in local economies, an understanding that will in turn allow us greater insight into early societies.

With this in the mind, Part II of this analysis will present the faunal background and data to which the preceding hypotheses concerning pig exploitation will be applied.

PART II

ZOOARCHAEOLOGY IN GREECE

Part II of this study presents the core of the zooarchaeological data on which this research is founded. It provides both the background for understanding why faunal analyses in Greece are vital to future research, and it sets out the raw data on which analyses in this study are founded.

Chapter 6, A History of Research and the Current State of Knowledge of Zooarchaeology in Greece, discusses the relevant background to faunal studies in Greece. The chapter addresses the treatment of archaeological animal bone assemblages in Greece, by first examining some of the reasons behind the current state of Greek faunal analysis. It discusses recent multidisciplinary work in Greek archaeology and summary works relating to fauna. There is brief discussion of the range of species typically found in Greek contexts, and this is followed by a look into major themes in zooarchaeology in Greece and a chronological summary of published sites with faunal analysis. The chapter serves as the background context within which to place the specific Early Helladic data analyzed in Chapter 7.

Chapter 7, The Early Helladic Faunal Assemblages at Helike, Tsoungiza, Lerna and Tiryns, contains the raw data on which this study is built. It presents each site under analysis, providing a brief description of the site, and an in-depth discussion of the faunal remains. Within each faunal section there is a discussion of the methodology employed by the analyst, followed by detailed examination of the assemblage. Where possible, analysis of each assemblage is broken down by species, and within these smaller categories, additional factors that have the potential to address economic aspects of the respective animal economies are addressed, such as utilization of secondary products, skeletal element frequencies and mortality profiles. Chapter 7 draws to a close with a summary discussion of the faunal data from Helike, Lerna, Tsoungiza and Tiryns, and what it suggests about Early Helladic social complexity in general. This sets the stage for the analysis of pig remains as a yardstick of economic complexity in ancient settlements, and thus a proxy indicator of social complexity – the subject of Part III of this study.

CHAPTER 6

A HISTORY OF RESEARCH AND THE CURRENT STATE OF KNOWLEDGE OF ZOOARCHAEOLOGY IN GREECE

In many respects, the study of zooarchaeology in Greece is no different than faunal studies in other regions of the world, while in other respects it differs drastically. Greece is similar to many other regions of the world in that it has a long history of intensive excavation dating back over a century. A wealth of sites span every temporal period from the Pleistocene through the modern age. The majority of these sites contain animal remains, and some have published faunal reports. I would argue that in this respect, zooarchaeology in Greece is not significantly different from zooarchaeology in Northern Europe, Mesoamerica or North America. Yet, zooarchaeology in Greece differs in a very large respect. Questions have not been asked regularly of its fauna.

Greek zooarchaeology has not been put to an interpretative use, and both the quantity and quality of analyses that have been done are highly varied. In addition, a significant problem is that faunal analyses in Greece have traditionally remained un-integrated with anthropological studies. This marked point of departure sets Greece apart from – and behind – other regions of the world in zooarchaeology, such as North America, the Near East, and perhaps Mesoamerica, in methodology and general intensity.[20] The lack of systematic and intensive faunal analysis in Greece is at once both frustrating and exciting. Frustration confronts anyone examining faunal reports in order to approach larger questions, such as inter and intra site socio-economic relationships. At the same time, zooarchaeology in Greece is exciting for the huge potential it holds in its ability to approach these questions. With few exceptions, (Halstead 1996, 2003; Halstead and Barrett 2004; Hamilakis 2003) zooarchaeology in Greece has not been used to address socio-economic questions.[21] Thus, faunal analyses in Greece are heavy with potential in the socio-economic arena.

This chapter examines the state of zooarchaeology in Greece by addressing these criticisms, and more, in the following way. It will begin by examining the importance of faunal studies to Greek archaeology and the potential avenues of research opened through their use. Next, it will inspect the reasons for the schismatic nature of faunal analysis in

---

[20] Of the 2,000+ faunal studies indexed for "animal remains" in learned journals by the Tozzer Library from 1980-2002, regions such as France, England and Germany have hundreds of entries, while Greece has a mere 30.

[21] Recent attention has been turned toward the role of Mycenaean feasting in the Late Bronze Age (Wright 2004, among others); however the social uses of faunal analysis in the Early Bronze Age remain unexplored.

Greece. This will be followed by a brief summary of important interdisciplinary projects over the past few decades that illustrate the growing methodological changes taking place in Greek archaeology. The major fauna, both wild and domestic, will be examined, followed by a discussion of the guiding historical methodologies in Greek zooarchaeology. The chapter will conclude with a review of the major archaeological site reports with published faunal analyses.

I. Potential Applications for Faunal Analysis in Greece

In 1826 when Brillat-Savarine wrote "Dis-moi ce que tu manges, je te dirai ce que tu es",[22] he may not have been referring to Early Bronze Age Greece, but his comment could not have been more à propos. Archaeology in Greece is only beginning to realize that the utility of faunal analyses reaches far beyond species lists and simple reconstructions of patterns of animal exploitation, which focus solely on dietary reconstruction.[23] The potential of faunal analyses as both an interpretive and integrative tool in the reconstruction of socio-economic aspects of complex societies has already been illustrated in the preceding chapters. Therefore, this discussion will focus on some of the areas of particular relevance to archaeology in Greece - areas which zooarchaeology has the greatest potential to augment.

Animal bones can be used to address issues of social stratification within a settlement, approaching differences in socio-economic status between households. This can be accomplished through examination of the types of fauna exploited, as well as the anatomical elements present and their relative frequency. Analysis of the faunal remains from seventeenth and eighteenth century cess-pits in Amsterdam established social (Jewish, non-Jewish) and economic (poor, average, rich) differences between households (Ijzereef 1988). The presence of certain species, in this case pig, allowed for the differentiation between Jewish and non-Jewish households. The spatial distribution of certain body parts permitted the economic status of the household to be determined, with skull parts, metapodia and phalanges (the less-desirable parts) that had been broken in many cases for marrow extraction, indicating a poorer household (Ijsereef 1988).

The utility of faunal analysis is especially pertinent to the exploration of social stratification in Early Helladic Greece. Architectural evidence clearly illustrates

---

[22] "Tell me what you eat, I'll tell you who you are."

[23] A more recent focus on the animal economy of Greece vis à vis secondary product exploitation is an exception to this statement (See Halstead 1996, 2003; Greenfield and Fowler 2003, 2005).

differentiation in the size of structures both intra- and inter-site in this period (Hägg and Konsola 1986). Does this differentiation suggest the presence of social hierarchies? Faunal analysis has the ability to address this question – and it is a question that forms a key area of inquiry in the study of social complexity.

Faunal assemblages can also be used to address economic complexity in respect to intra- and inter-site economic relationships involving the exchange, redistribution and trade of animals and animal products (Clark 1987). Examining age profiles and sex ratios of domestic animals enables us to draw conclusions as to the use of particular animals for primary products, or as secondary product producers. A faunal assemblage dominated by older adult females may indicate a site at which that particular species, sheep for example, were exploited primarily for their milk or wool. Though discussion has taken place on this particular use of fauna from archaeological assemblages, including studies from the Greek Neolithic (Halstead 1981) and Bronze Age (Pullen 1992), its potential has been limited in scope to an emphasis on secondary products. In Greece, little is known about inter-site economic relationships during the Early Helladic period. Faunal analyses that focus on the utilization of pigs as an indicator of economic complexity have the potential to change this. Additionally, harvest profiles and their relationship to different economic strategies are an approach overlooked in Greek contexts, and can be addressed by examining skeletal element frequency, age profiles and sex ratios of faunal assemblages from Early Helladic settlements.

Hunting practices and myriad related issues associated with the hunt are also important in Greek faunal analyses. Animal bones can inform us about hunting practices through an examination of the percentage of wild versus domestic fauna in an assemblage. This may provide information not only about subsistence and risk-buffering strategies, but also about power, prestige and social hierarchy. Hunting in Mycenaean society provides a good example of this interpretive use of fauna, as hunting wild animals is one of the most common themes in Mycenaean iconography (Hamilakis 2003). Hunting in ancient society is also an excellent example of the utility of integrating faunal analyses with other aspects of the site, such as iconography in the form of painting and figurines. Independent analyses of the fauna and the remaining small finds of an assemblage can be combined to obtain a fuller, more complete picture of the practices under analysis. This type of analysis is perhaps more applicable to later chronological periods than the one addressed in this study, but still its potential utility for Greek archaeology should be acknowledged.

The preceding examples demonstrate just some of the *potential* applications of zooarchaeology. A further shortfall in Greek faunal analyses is the failure to integrate the fauna with the rest of the site. Though some studies have used fauna to address economic questions,[24] these questions, as already mentioned, have largely centered on the production of secondary products from sheep and goat husbandry. Faunal analysis has also been used in discussion of the timing of the domestication of a limited number of species,[25] however many of these examinations still stop short of integration. When addressing questions of the economic importance and significance of animals, we should be seeking to understand the relationship between humans and animals, including how animal products are provided at each site. Do specialist producers, such as nomadic pastoralists or specialist hunters, provide animal products? Are they locally managed in the care of individual families? Can we identify consumer sites and producer sites, or are producers and consumers located within the same site? How is meat made available, distributed, consumed? Perhaps most importantly, what is the relationship between rural and urban sites? This last question is especially pertinent to Early Helladic Greece.

Applying faunal analyses to the study of the socio-economic relationship between large administrative settlements and small villages is an endeavor that could shed much-needed light on the daily economies of small-scale rural settlements. Knowledge of the outlying villages in prehistoric and historic Greece is nearly non-existent. How can we truly understand the Greek Bronze Age, or the Classical period for that matter, without understanding daily life in the rural villages? We could begin to understand small-scale subsistence economies and their social organization by examining the fauna present at these sites, determining their function, and then comparing them to the fauna from the larger settlements. Such a comparison could reveal where animal products were produced, elucidate trade networks, and provide insight into the economic relationships between large and small-scale settlements. It might also be able to address questions surrounding differences in social status based upon the relative distributions of skeletal part frequencies and taxa. In addition, integration of the fauna with paeleoethnobotanical studies could help reconstruct climate and provide a more complete picture of periods in question. We can even compare the diets of an animal, such as pigs, to the diets of humans (via isotope analysis), in order to address the role of that animal in a settlement. For example, if we find the diets of

[24] See Greenfield and Fowler 2003; Halstead 1981, 1996, 2003

[25] Such discussions are usually limited to sheep, goats, cattle and occasionally pigs. See Gejvall 1969; Jones 1987; Mancz 1989

pigs and humans to be similar, we might infer that pigs were kept around the settlement. This may shed light on the management of the animal, as well as the inter-relationship between pigs and humans.

Additionally, animal bones have the potential to inform us about ritual practices. There is a developing body of research on the role of sanctuaries and feasting in the Late Bronze Age of Greece,[26] which, if combined with faunal analysis could hold great potential. In the past, studies of this nature have focused more on the identification of ritual and its description, with little to no emphasis placed on the social implications of sacrifice (Hamilakis and Konsolaki 2004). An integrative and interpretive examination of the fauna could change this. Zooarchaeology holds great potential in Greece, enabling researchers to fully explore many aspects of ancient social, economic and human-animal relationships that heretofore have only garnered attention in respect to textual references and artistic representations.

II. Reasons behind the Current State of Greek Faunal Analysis

The wealth of possibility offered by zooarchaeology raises the question why has faunal analysis in Greece often been overlooked on so many excavations? Reasons for the schismatic nature of faunal studies in Greece vary. A realistic explanation of the sidelining of faunal studies in Greece is arguably the attention that classical archaeologists, focused on the acquisition of museum worthy artifacts, have been prepared to spend on them. Archaeological sites in Greece, especially in later historical periods, contain an abundance of other material, such as pottery, monumental architecture and household artifacts. Faunal analyses have faired better in prehistoric periods, if only because of the relative paucity of other types of material remains. The wealth of textual information in Greece has thus been both a boon and bane to anthropological archaeologists. Although methods employed in the field of Greek archaeology in general have come under scrutiny during the last few decades, scholarship in this area often still exhibits…

> …an insularity fostered by the conventions of anchoring field research to classical texts. While no one can deny the many contributions such text-aided archaeology produced, the other side of the coin was a reluctance to adopt many of the new methods and theoretical frameworks current in North America and Europe (Kardulias 1994:1).

[26] See Halstead 2003; Wright 2004a.

Zooarchaeology in Greece continues to be a slave to chronology. Archaeologists working with the remains of more socially complex societies tend to be confronted with assemblages containing an abundance of inorganic artifacts, such as art, architecture, and in many cases texts. In Greece, Late Bronze Age and later sites are more likely to contain a wealth of material remains outside the faunal realm, and as such, the attention of archaeologists tends to be focused elsewhere, thereby overlooking the animal remains or relegating them to an appendix. A problem highly applicable to faunal analyses in Greece is that…

> …bones seem to be regarded as a fall-back source of archaeological information, to be interpreted in an archaeological context when there is little other evidence, but increasingly sidelined when there is a substantial corpus of (other) artefactual or structural data (O'Connor 1996:11).

An additional, rather large problem is that faunal remains are still, to this day, not regularly collected from historic sites in Greece, rendering the record quite poor due to a lack of the recognition of the importance of animal bone. An example of this methodological oversight can be read in the American School of Classical Studies' archaeological site manual to the Corinth Excavations from 2002. In regard to the excavation and recovery of fauna, it reads, "Only keep bones from levels that provide a good sample" (Sec 2.5:6). Nowhere, however, is "good sample" defined. When animal bones are collected, problems still arise. Those examining faunal assemblages from Greek contexts are confronted with highly variable collections.

> The samples differ as to size because of the number of years of excavation, the number of years of study of the fauna (some reports are only on part of the sample excavated) and because of the recovery methods used (hand-collection, dry-sieving, water-sieving, floatation) (Reese 1994a:192).

I hope that a more complete picture has now emerged explaining the fractured state of zooarchaeology in Greece. The importance and potential of faunal analysis has been severely overlooked in Greece because it was traditionally obscured by text and artifact driven approaches that focused on spectacular palatial, administrative and sanctuary sites. This is still a problem facing zooarchaeology in Greece, but rather than allowing this critique to strike fear in the minds of those studying fauna in Greece, it can serve as background information that allows one to understand the framework from within which archaeology in Greece was, and often still is, practiced. Moreover, this critique can also provide an understanding of the great potential faunal analysis holds. Since, until quite recently, archaeology in Greece was dominated by concern with description and classification, and limited or lacking in "creative approaches to the interpretation of the archaeological remains in human terms" (Kardulias 1994:2); archaeology in Greece has not had as its focus an understanding of what types of social or economic information faunal remains can provide. This opens the door to a wealth of potential for very significant and vital research in Greek archaeology.

Rarely have faunal analyses been integrated within the larger context of the site in Greece. A great irony lies in the relative abundance of faunal remains in most archaeological contexts, and the failure to recognize their potential in addressing socio-economic issues beyond subsistence. Plant and animal remains have been recovered on most sites in Greece – and yet, at the vast majority of sites, they are accorded little more than cursory mention. This fact is beginning to garner increasing recognition in the current theoretical archaeological literature (Gamble 2003; Halstead 2003; Hamilakis 2003; Kardulias 1994; Kotjabopoulou et al. 2003; Payne 1985; Pullen 1992; Reese 1994a., among others). Future work must not only put an effort into adequately recovering the faunal remains, but it must then focus on analysis and integration. Zooarchaeology in Greece must begin to ask what the fauna can tell us about larger issues, such reconstructions of ancient climate, subsistence strategies, small-scale local economies, inter-site relationships, seasonal exploitation of land and resources, pastoralism, herding, and perhaps even the geographic movement of peoples.

III. Recent Multidisciplinary Work in Greek Archaeology

Lest it appear that the entire field of Greek archaeology is mired in antiquated ideas and armchair-bound, pipe-smoking dilettantes, efforts have been made to bring change to this classical, descriptive tradition. Several expeditions have carried out systematic regional surveys in Greece over the past few decades, and their success serves to illustrate the potential offered by multi and inter-disciplinary approaches carried out within a systematic, scientific framework. Since these surveys did not focus directly on faunal analyses, they will be but briefly noted in this paper as examples of the turning tide in Greek archaeology.

The Minnesota Messenia Expedition (MME) was the first of these pioneering studies, whose impetus originated as an attempt to place the excavations at Pylos into a broader environmental context (McDonald and Rapp 1972). The Southern Argolid Survey (Jameson et al. 1994) derived its impetus from a similar goal, surrounding excavations at Franchthi Cave and classical Halieis. Other interdisciplinary projects include the Cambridge and Bradford Boeotia Archaeological and Geological Expedition (Bintliff and Snodgrass 1985), the Grevana Project (Savina et al. 1991), The Nemea Valley Project (Cherry et al. 1988) and the extensive work by the American School of Classical Studies at Athens in the area surrounding Ancient Corinth (Eastern Corinth Archaeological Survey). Current research still continues in this vein with the 2004 volume, edited by Alcock and Cherry (2004) of collected research critical of the use to which these surveys have been put. These projects illustrate a paradigmatic shift that is attempting to assert itself in Greek archaeology. This shift is also beginning to appear in the literature on faunal analyses.

IV. Summary Works on Faunal Analysis in Greek Archaeology

The first undertaking of a synthesis of faunal analysis in archaeological sites in Greece was not attempted until Sebastian Payne's 1984 publication "Zooarchaeology in Greece: A Reader's Guide." Until recently, scant attention was accorded to the animal remains from archaeological contexts. Few studies placed any reliance on zooarchaeological evidence beyond establishing the presence of domestic animals and drawing quantitative analyses focusing upon NISP and MNI/MNE ratios (Halstead 1996). In this respect, the state of zooarchaeology in Greece is today very much in its infancy. It is all too common practice in the analysis of site assemblages to treat animal bone as either a separate entity, neglecting to integrate the faunal remains with the larger site interpretations, or to relegate them to a table summarizing numbers and types of extant fauna in the back of a volume. Zooarchaeology in Greece is decades behind faunal analyses in other regions, such as the rest of Europe, the Near East and North America, still failing to integrate faunal studies with larger socio-economic lines of questioning.

Ten years after the appearance of Payne's publication, David Reese (1994a.) built upon Payne's assessment, contributing to the synthesis of zooarchaeology in Greece with an article entitled "Recent Work in Greek Zooarchaeology" in which he summarized recent contributions to the body of faunal knowledge.[27] Reese compiled papers appearing after

[27] David Reese has compiled an update to these references which he has graciously made available to me, covering material through 2004. Examination of these works still yields few applications of zooarchaeology to larger questions of social organization, with those that do pertaining to the Late Bronze Age/Mycenaean periods).

1983, including articles dealing with invertebrates (an omission of Payne's) and bone reports not included by Payne. Reese departed significantly from Payne's work in that he included only post-Bronze Age sites. Reese (1994) argued that the main body of material on faunal analysis in Greece was produced from study on Neolithic and (Late) Bronze Age sites, and therefore chose to concentrate on later periods in order to provide a balance to the nature of the known record. By omitting faunal evidence prior to the Bronze Age, Reese's approach overlooked significant contributions to the field of zooarchaeology in Greece.[28] Thus, while a useful research tool for those interested in later periods, it is an incomplete picture of "recent" work Greek faunal analysis.

The most recent major published work on zooarchaeology in Greece marks a significant departure from the two previously discussed summary papers. *Zooarchaeology in Greece: Recent Advances*, published in 2003 is a compilation of research focusing on faunal studies and their utility in Greek archaeological contexts (Kotjabopoulou et al. 2003). Adopting a novel approach in Greek faunal analysis, this volume examines animal bones within wider methodological contexts, such as the integration of zooarchaeology with ritual and intra-site economics. The volume does not focus on any specific site or time period, but instead attempts to illustrate how faunal analysis can contribute to a greater understanding of archaeological assemblages. This is a milestone in Greek zooarchaeology, standing as a marker of the kinds of questions faunal analyses can approach.

Although recent work in zooarchaeology in Greece is encouraging, it does not provide an in depth understanding of the current state of faunal analysis in Greece. Such an understanding can be obtained best through an examination of a sample of published volumes on the fauna from Greek sites. Given the uneven nature of research on faunal assemblages from Greek archaeological contexts, it is beneficial to concentrate on the production of a brief overview of known sites spanning the entirety of the country geographically and diachronically. In advance of a discussion of faunal reports, however, a survey of the species of fauna, both wild and domestic, that have been commonly recovered from archaeological contexts in Greece should be made. This faunal review will then be followed by a survey of

[28] Many recent analyses in Greek zooarchaeology have been carried out on Paleolithic assemblages. Of particular note is the extensive work at Klithi (Gamble 1997). Additionally, faunal analysis is being utilized in Macedonia, at Megalo Nisi Galanis (Greenfield and Fowler 2005), to address issues surrounding the secondary products revolution during the Late Neolithic/Early Bronze Age. While of great importance, Macedonia differs significantly in its material culture, geography and climate to warrant exclusion from discussions of EH society in the Peloponnese. It can, and should, however, serve as a model of the use to which faunal analyses should be put.

a sample of the extant published material on fauna from Greek archaeological assemblages, accompanied by a summary of the types of analyses performed on these data.

V. The Fauna

Thus far, this examination has addressed the history of zooarchaeology in Greece and potential uses for faunal analyses. However, a more complete picture of zooarchaeology in Greece should also incorporate mention of the species that were extant, as well as an overview of the major animal domesticates present.

Evidence for Indigenous Fauna:

The Mesolithic strata of Franchthi Cave in the southeast Argolid in the eastern Peloponnese offers useful information on the types of animals that were present in Greece, or at least in this region of Greece, before domestication. In these strata red deer (*Cervus elaphus*) bones represent the most common large mammal present (Payne 1982; Perles 2001). Wild boar (*Sus scrofa*), wild cattle/auroch (*Bos primigenius*), roe deer (*Capreolus capreolus*), fox (*Vulpes vulpes*), hare (*Lepus europaeus*), badger (*Meles meles*), wildcat (*Felis silvestris*), beech martin (*Martes foina*), as well as some bird and fish remains are also present. Finds from Paleolithic/Mesolithic strata at Theopetra Cave add bear (*Ursus arctos.*) and hyena (*Crocuta* sp.) to the list of indigenous species (Newton 2003). The Paleolithic site Klithi, in northwestern Greece yields similar data, with a few additional species (Gamble 1997). Fauna at Klithi include those found at Franchthi, as well as beaver (*Castor fiber*), lynx (*Felis lynx*), wolf (*Canis lupus*), ibex (*Capra ibex*) and chamois (*Rubicapra rubicapra*). Wild ass (*Equus hydruntinus*) is present at Klisoura Cave (Koumouzelis et al. 2001). In addition to the mammals from Franchthi Cave, evidence from Neolithic sites in Thessaly and Macedonia, both in the northern part of the country, yield evidence for wild goat (*Capra hircus*) exploitation (Bokonyi 1986; Larje 1987). Additional faunal evidence from Makri (Curci and Tagliacozzo 2003) contains evidence for the exploitation of dog (*Canis familiaris*) and fallow deer (*Dama dama*). Lion (*Panthera leo*) bones have also been recovered (Yannouli 2003) (*Table 6.1*). Wild mammals continued to be exploited in the Neolithic, and their continued presence during this period in many assemblages provides a good picture of many of the types of indigenous fauna available for exploitation in prehistoric Greece.[29]

[29] While many factors may affect the recovery of animal bone, evidence for the exploitation of similar species exists, spanning both geographic regions and time. And although the species mentioned should not be

Red deer (*Cervus elaphus*)	**Cave hyena** (*Crocuta spelaea*)
Roe deer (*Capreolus capreolus*)	**Beaver** (*Castor fiber*)
Fallow deer (*Dama dama*)	**Dog** (*Canis familiaris*)
Wild boar (*Sus scrofa*)	**Wolf** (*Canis lupus*)
Auroch (*Bos primigenius*)	**Ibex** (*Capra ibex*)
Fox (*Vulpes vulpes*)	**Chamois** (*Rubicapra rubicapra*)
Hare (*Lepus europaeus*)	**Goat** (*Capra hircus*)
Wildcat (*Felis sylvestris*)	**Badger** (*Meles meles*)
Lynx (*Felis lynx*)	**Beech martin** (*Martes foina*)
Lion (*Panthera leo*)	**Wild ass** (*Equus hydruntinus*)

Table 6.1 **Summary of indigenous fauna recovered from prehistoric assemblages**

Evidence for Early Animal Domesticates:

Excavation of Neolithic strata at sites across Greece yields very similar lists of exploited species (Boessneck 1962; Bokonyi 1986; Gevjall 1969; Halstead and Jones 1980; Jarman and Jarman 1968). These include sheep (*Ovis aries*), goat (*Capra hircus*), auroch/large bovid (*Bos primigenus?*), pig (*Sus scrofa*), horse/donkey (Equus sp.) and dog (*Canis familiaris*) (Perles 2001). While these faunal reports provide information on the types of species present, they should be used critically in discussions of species distribution and abundance. As previously mentioned, the uneven nature of excavation and recovery methods in Greek archaeology must be given serious consideration when drawing large conclusions about fauna in Greece, especially in regard to the exploitation of wild and domestic animals. For example it is certainly possible that the importance of wild resources might be underestimated due to poor preservation and recovery, or even vice versa (Halstead 1989). Furthermore, the presence of wild resources in an assemblage does not necessarily imply their widespread exploitation. A large portion of subsistence economy may be culturally based and must be "studied as the expression of social choices within the possibilities offered by the environment and the level of technical development. Which species are exploited, and how, depends as much on traditions and ideology as on economic considerations" (Perles 2001:152).

Given these, and other methodological considerations, there are still valid generalizations about the nature and character of early animal domesticates in Greece that can be drawn. For example, sheep and goat are often the most common species represented in faunal assemblages. Either cattle or pig bones, depending upon the site, and the time period,

interpreted as a complete list of all wild fauna present in Greece before domestication, they can be interpreted as evidence of some of the types of species present and exploited.

follow sheep and goat. Small quantities of dog bones round out faunal assemblages, which after the early Bronze Age also include donkey and horse. More involved questions concerning the frequencies and interpretations of these fauna are questions that have been addressed in differential intensities in many of the published reports. A discussion of some of these reports will now follow, accompanied by summary discussion of the types of questions zooarchaeology in Greece has been used to address. This analysis will provide a fuller picture of the state of faunal analysis in Greece.

VI. Major Themes in Greek Zooarchaeology and a Summary of Published Sites Containing Faunal Analyses

The first report to provide detailed information on faunal remains from an archaeological site in Greece was published in 1956, with Joachim Boessneck's report on the animal bones from two prehistoric sites in Thessaly, Arapi-Magula and Otzaki-Magula (Boessneck 1956; Payne 1985). However, the first bone studies performed in the Aegean region by zoologists were done in 1880 for Heinrich Schliemann on the faunal remains from Troy (Schliemann, 1880). And the first such study actually published by a zoologist was on a small sample of around 50 bones from the 1901 excavation in the Late Minoan III and Early Iron Age Dictaean cave on Crete (Boyd-Dawkins 1902; Reese 1994a.). Boessneck's seminal 1956 work was then a notable first attempt at faunal analysis. These past fifty years, however, have witnessed a marked rise in the attention accorded to animal bones. The majority of this attention has focused on issues of domestication, not the attendant socio-economic questions faunal remains may be able to answer. Although we have seen progress in the recognition of the merits of studying faunal assemblages, many archaeological reports still continue to treat animal bone in isolation from the rest of the site.

There have been several analyses of fauna from Greek assemblages since Boessneck's first published work in Thessaly (Boessneck 1956) that are worthy of note for their pioneering efforts. Zooarchaeological studies have been conducted on assemblages from nearly every region of Greece. These published faunal assemblages vary greatly in size, quality of retrieval, analytical methods and details of presentation (Halstead 1996). The following is a brief account of some of the better-known published sites in Greece containing mention and/or analysis of faunal remains. This list is by no means exhaustive, as a complete catalogue of every site in Greece with faunal remains is a large task beyond the scope of this current study. The aim of this chapter is to illustrate that given the abundance of sites with faunal remains, only a few have been systematically studied. It is worth noting that this trend is changing, and I will address current analytical trends in Greek zooarchaeology shortly. The following, then, is a summary discussion of a sample of published faunal material from Greek contexts, which illustrates the nature of the extant faunal evidence in Greece diachronically.

Paleolithic & Mesolithic Sites:

There are a handful of Paleolithic sites with published fauna in Greece. The most extensively excavated of these are the following: Petralona (Kretzoi and Poulianos 1981) in Macedonia; Klithi (Gamble 1997), Perama Cave (Pavlakis et al. 2003), Kastritsa (Higgs et al. 1967) and Asprochaliko (Bailey et al. 1983) in Epirus; Franchthi Cave (Payne 1982) in the Peloponnese; Grava on Corfu (Sordinas 1969), Kitsos (Jullien 1973) and Seidi (Schmid 1965) in central Greece, and Peneios Valley (Boessneck 1965) in Thessaly. Several Mesolithic sites show continuous use from the Paleolithic - Theopetra Cave in Thessaly (Karkanas 1999; Newton 2003), Franchthi Cave and Klisoura Cave in the Peloponnese, and Asprochaliko in Epirus. The quality and scholarship of these faunal reports varies widely, illustrating the uneven nature of faunal analysis in Greece (*Table 6.2*). Analyses of Franchthi Cave have attempted to use the fauna in dating, as well as ancient climactic and environmental reconstructions, a rare (in Greece) and wonderful interpretative use of fauna (Payne 1982). The fauna from Klithi have also been analyzed in an interpretative context, with focus on the taphonomic processes that affected the animal bone assemblage and attempts to place Klithi within the wider context of Late Glacial animal bone studies (Gamble 1997). Recent work in Theopetra Cave has focused on examining mineral assemblages to understand diagenesis (Karkanas et al. 1999).

Some faunal reports provide detailed analysis of a particular species, such as the focus on carnivores at Petralona (Kretzoi and Poulianos 1981). Other published faunal reports are very limited in scope, providing little else aside from species lists. These sites include Asprochaliko, Kastritsa, Grava, Kitsos, Peneios Valley, and Seidi (*Table 6.2*).

SITE	*LOCATION*	*ANALYSES PERFORMED*[30]
Petralona	Macedonia	SL; M; A; DR (carnivores; *Dicerorhinus*)
Franchthi Cave	Peloponnese	SL; D (climatic reconstruction)
Klisoura Cave	Peloponnese	SL; NISP; MNI; D
Kastritsa	Epirus	SL; M (*Bos* only)
Perama Cave	Epirus	SL; M; D
Asprochaliko	Epirus	SL
Klithi	Epirus	SL; S; NISP; A; M; D
Kitsos	Central Greece	SL
Seidi	Central Greece	SL
Peneios Valley	Thessaly	SL; S; M
Theopetra Cave	Thessaly	SL; D
Grava	Corfu	SL

Table 6.2 **Faunal analyses at Paleolithic/Mesolithic sites (see text for references)**

These representative faunal reports of Greek Paleolithic assemblages demonstrate great variability in the intensity of analysis. Earlier reports tend to be composed of simple species lists, while more recent analyses focus on placing the fauna within a larger, interpretative context, typically addressing environmental and subsistence-related issues. Thus in Paleolithic and Mesolithic assemblages, the trend is to use animal bones in intra-site analyses, such as subsistence practices, butchery techniques and the range of species exploited. This application of fauna to larger questions may be due to the intrinsic nature of prehistoric sites in chronological periods in which there are few material remains with which to work.

Neolithic Sites:

The last few decades have seen a great deal of interest in the Neolithic. The Neolithic in Greece has received attention for the beginnings of animal husbandry and domestication of crops and animals. It is also a prehistoric period in which there are no large palatial complexes, nor socially stratified societies. For these reasons, research in the Greek Neolithic tends to be more systematic, being in the purview of anthropologists and paleontologists. Researchers in the Neolithic tend to have a greater focus on both faunal and floral analyses, areas largely neglected by archaeologists working in later historical periods.

[30] SL = species list S = skeletal element frequency A = age determination
DR = detailed report on specific species M = measurements D = discussion/analysis
NISP = number of identified specimens MNI = minimum number individuals

Among those Neolithic sites with published treatment of animal bones are Sitagroi (Bokonyi 1986), Megalo-Nisi Galanis (Greenfield and Fowler 2003, 2005), Nea Nikomedeia (Higgs 1962), and Servia (Watson 1979) in Macedonia; Sesklo (Schwartz 1982), Achilleion (Bokonyi 1989), Otzaki-Magula (Boessneck 1956), Agia Sofia-Magula (von den Driesch and Enderle 1976), Arapi-Magula (Boessneck 1956), Argissa-Magula (Boessneck 1962), Dimini (Hourmouziadis 1979; Halstead 1992), and Prodromos (Halstead and Jones 1980) in Thessaly; Makri (Curci and Tagliacozzo 1998) in northeastern Greece; Paradeisos (Larje 1987) in northern Greece; Kitsos (Jullien 1973) in central Greece; Franchthi Cave (Payne 1982) and Lerna (Gejvall 1969) in the Peloponnese; Knossos (Jarman and Jarman 1968)

SITE	*LOCATION*	*ANALYSES PERFORMED*[31]
Sitagroi	Macedonia	SL; A; S; M; D; DR (*Dama*)
Megalo Nisi Galanis	Macedonia	SL; NISP; A; S; M; D; DR
Nea Nikomedeia	Macedonia	SL; A; M
Servia	Macedonia	SL
Achilleion	Thessaly	SL; M; S; D
Otzaki-Magula	Thessaly	SL; M; S; A (*Sus*)
Agia Sofia-Magula	Thessaly	SL; M; A; S; D
Arapi-Magula	Thessaly	SL; M; A (*Sus*); S; D
Argissa-Magula	Thessaly	SL; M; A; S; D
Prodromos	Thessaly	SL; M; A; S; D
Dimini	Thessaly	SL; M; A; S; D
Sesklo	Thessaly	SL; M; A; S; D
Makri	Northeastern Greece	SL; NISP; MNI; A; D
Paradeisos	Northern Greece	SL; M; A; S; D;
Kitsos	Central Greece	SL; DR (fish)
Franchthi Cave	Peloponnese	SL
Lerna	Peloponnese	SL; M; S; A; D
Knossos	Crete	SL; D
Phaistos	Crete	SL
Aghios Petros	Sporades	SL
Emporio	Chios	SL; M; S; A
Kephala	Kea	SL; M; S; A
Saliagos	Antiparos	SL; M; A; DR (*Dama*; fish)

Table 6.3 **Faunal analyses at Neolithic sites (see text for references)**

[31] SL = species list S = skeletal element frequency A = age determination
DR = detailed report on specific species M = measurements D = discussion/analysis
NISP = number of identified specimens MNI = minimum number individuals

and Phaistos (Pernier 1935) on Crete; Aghios Petros (Swartz 1982) in the Sporades, Emporio (Clutton-Brock 1982) on Chios, Kephala (Coy 1977) on Kea, and Saliagos (Higgs et al. 1968) near Antiparos (*Table 6.3*).

Faunal analysis on Neolithic assemblages in Greece experiences similar differences in quantity and quality of analysis as analyses in the Paleolithic. The largest difference arises in some reports' analysis and mention of animal husbandry and domestication. As previously noted, the Neolithic is the chronological period that yields (arguably) the first evidence of animal domestication in Greece, therefore faunal reports on Neolithic assemblages frequently contain at the very least, species lists of wild and domestic fauna. Many published works only consist of small species lists,[32] while others address basic issues of subsistence by looking at skeletal part frequencies,[33] age profiles[34] and sex ratios - although it is still rare to find such extensive treatment of fauna. The exploitation of secondary products also receives attention, especially in attempts to reconstruct ancient economies (Greenfield and Fowler 2003, 2005; Halstead 1981, 1992, 2003; Sheratt 1981), as do questions surrounding pastoralism and household herding (Halstead 1996). Although fauna are beginning to be employed as tools necessary to understanding prehistoric subsistence and economy (Curci and Tagliacozzo 2003), many such discussions do not occur in the site reports themselves, but rather in the secondary literature of journal forums.[35] While discourse, such articles tend to have narrow foci on one particular issue. They do not provide complete pictures of sites, nor do they frequently integrate their analysis with the wider context of the Neolithic as a larger entity.

In addition to inconsistency in faunal analysis, the geographic distribution of sites is also problematic. At present, excavated Neolithic sites in Greece are geographically concentrated in Thessaly and Macedonia, with a smattering on many of the islands. Excavation bias may help partially explain this distribution, as these are regions that have been the focus of extensive fieldwork spanning many decades. Both Thessaly and Macedonia have a history of intensive field projects, and as in most parts of Greece, there is continuity of use and habitation through time at many, if not most of these sites. Continuous occupation is noted in nearly every published site report. Thus places like Sitagroi, Argissa-

[32] See Servia (Watson 1979), Sesklo (Schwartz 1982) and Phaistos (Pernier 1935) as common examples.
[33] See Agia Sofia-Magula (von den Driesch and Enderle 1976), Argissa-Magula (Boessneck 1962), Lerna (Gejvall 1969), Pefkakia (Jordan 1975), and Sitagroi (Bokonyi 1986).
[34] See Agia Sofia-Magula (von den Driesch and Enderle 1976), Lerna (Gejvall 1969) and Pefkakia (Jordan (1975).
[35] Greenfield and Fowler's 2005 volume on Megalo Nisi Galanis is a noted, very recent, exception.

Magula, Knossos, Franchthi Cave and Klithi have yielded evidence for occupation and use in more than one epoch. Larger excavations, however, tend focus on those periods of the greatest interest to the excavator, in many cases overlooking periods of lesser personal interest.[36]

Bronze Age Sites:

The Bronze Age in Greece has been the focus of much scholarly attention in light of increasing social complexity, palatial settlements, and of course the Linear B tablets. Monumental excavations at Lerna, Mycenae, Tiryns, Nichoria, Pylos and Knossos have all provided excellent sources of information on increasing social complexity in Greece. Faunal reports comprise a good portion of the published reports; however they are again often relegated to quantified lists of extant species. When analysis of faunal remains has been undertaken, focus has centered on the domestication and utilization of sheep and goats, with emphasis accorded to issues of the "secondary products revolution"[37] and the exploitation of domestic fauna (Cosmopoulos et al. 2003; Halstead 1981, 1986, 1987; Pullen 1992). While such study is much preferred to the faunal lists generated at most sites, its treatment has been exhausted. We know from administrative Linear B evidence of the importance of sheep and goat husbandry in the palatial economy (Halstead 2003), and we can confirm this importance with a glance at the relative numbers of ovicaprine remains in the faunal record. But as of yet, we have still neglected to examine the greater socio-economic implications of the other fauna. In addition, much of the focus of archaeological investigation in Bronze Age Greece has centered on the large-scale palatial and administrative centers such as Mycenae, Tiryns, Pylos and Lerna. These extensively excavated sites each contain a wealth of animal bone; however the faunal assemblage at Tiryns is the only one from an administrative context to have been published in detail (von den Driesch and Boessneck 1990). The fauna from Lerna have been published, and are currently being restudied by David Reese. In addition, faunal remains from Tsoungiza (Nemea) have been analyzed by Paul Halstead and are currently in press (Pullen 2006, in press). The eventual dissemination of the Tsoungiza material will hopefully serve as a model illustrating the necessity of faunal analyses in Greek

---

[36] A good example of this trend is the well-published work on the Athenian Agora and Corinth by the American School of Classical Studies in Athens, in which excavators have allotted the majority of their resources to work on the Classical and later periods. Earlier phases of occupation, such as the Neolithic, receive a brief comment and little further attention – at least at this moment in time

[37] The secondary products revolution has at its core the production of wool and dairy products for trade, ushering in increased economic complexity both within and between sites by promoting division of labor and potentially social stratification.

archaeological contexts. Thus, although faunal remains of the Bronze Age in Greece have received attention, there is room for more investigation concerning the smaller village economies and local social organization. Of great benefit would be comparative analyses between these small-scale settlements and the larger administrative centers. A zooarchaeological approach to such issues would be novel, potentially providing a wealth of information concerning the inter-site relationships between large and small settlements.

Numerous site reports contain reference to Bronze Age faunal remains in Greece. Still more are currently in excavation and awaiting publication.[38] Of note are the following:

SITE	*LOCATION*	*ANALYSES PERFORMED*[39]
Lerna	Peloponnese	SL; M; S; A; D
Nichoria	Peloponnese	SL; S; MNI; A (*ovicaprids, equids*); D
Midea	Peloponnese	SL
Tiryns	Peloponnese	SL; M
Pylos	Peloponnese	SL
Pefkakia	Thessaly	SL; M; A; S; D
Argissa-Magula	Thessaly	SL; M; A; S; D
Eleusis	Attica	SL; M; A; NISP; D
Sitagroi	Macedonia	SL; A; S; M; D; DR (*Dama*)
Pentapolis	Macedonia	SL; S
Kastanas	Macedonia	SL; M; S
Servia	Macedonia	SL
Assiros	Macedonia	SL
Agios Stephanos	Lakonia	SL
Akrotiri	Thera	SL; D
Knossos	Crete	SL
Myrtos	Crete	SL; D
Phaistos	Crete	SL
Dictean Cave	Crete	SL
Emporio	Chios	SL; M; A; S
Phylakopi	Melos	SL; S (*ovicaprids*); D

Table 6.4 **Faunal analyses at Bronze Age sites (see text for references)**

[38] Specifically, the fauna from FN-EH Tsoungiza at Nemea by Paul Halstead (Pullen 2006, in press)
[39] SL = species list S = skeletal element frequency A = age determination
DR = detailed report on specific species M = measurements D = discussion/analysis
NISP = number of identified specimens MNI = minimum number individuals

Lerna (Gejvall 1969), Nichoria (Sloan and Duncan 1978, Mancz 1989), Tiryns (von den Driesch and Boessneck 1990), Pylos (Nobis 1993), and Midea (Gejvall 1983) in the Peloponnese; Akrotiri (Gamble 1978) on Thera; Knossos (see numerous publications by Halstead and Gamble), Phaistos (Pernier 1935), Myrtos (Gamble 1979) and the Dictaean Cave (Boyd-Dawkins 1902, Reese 1994a.) on Crete; Pevkakia (Jordan 1975), and Argissa-Magula (Boessneck 1962) in Thessaly; Sitagroi (Bokonyi 1986), Pentapolis (Koufos 1981), Kastanas (Reichstein 1982), Servia (Watson 1979) and Assiros (Halstead and Jones 1980) in Macedonia; Agios Stephanos (Reese 1994) in Lakonia; Eleusis (Cosmopoulos et al. 2003) in Attica; Emporio on Chios (Clutton-Brock 1982) and Phylakopi (Gamble 1979, 1980) on Melos (*Table 6.4*).

The geographic distribution of Bronze Age sites in Greece with faunal components shows a marked concentration in Crete and in the Peloponnese. Differential excavation is certainly a key factor in this distribution. Both the eastern Peloponnese and Crete were sites of later Bronze Age palatial complexes and have been accorded rather heavy analysis. This preference for excavating remains of the Mycenaean kingdom accounts for the wealth of published material in these areas. Closer analysis of the nature of the excavated material in the Peloponnese and on Crete indicates that there is a paucity of material with a focus on smaller settlements, and thus a great amount of potential in exploring non-palatial local economies, social organization and subsistence strategies. The great majority of the aforementioned Bronze Age sites are Late Bronze in date, and thus there is a significant gap in knowledge of the Early Bronze Age.

Iron Age Sites:

Knowledge of the Iron Age in Greece is rather poor, and as one might expect, the availability of published material concerning this time period tends to be relegated to small portions of larger work, such as at Knossos on Crete. Other published references with mention of Iron Age fauna include Assiros (Halstead and Jones 1980) and Kastanas (Becker 1986) in Macedonia; Nichoria (Sloan and Duncan 1978; Mancz 1989) in the Peloponnese; Heraion (Boessneck and von den Driesch 1981) on Samos and Karphi on Crete (The British School of Archaeology at Athens) (*Table 6.5*).

SITE	LOCATION	ANALYSES PERFORMED[40]
Knossos	Crete	SL
Karphi	Crete	SL
Assiros	Macedonia	SL
Kastanas	Macedonia	SL
Nichoria	Peloponnese	SL; S; MNI; A (*ovicaprids, equids*); D
Heraion	Samos	SL

Table 6.5 **Faunal analyses at Iron Age sites (see text for references)**

Classical and Hellenistic Sites:

One might expect a significant volume of material to have been published from archaeological sites dating to the Classical and Hellenistic periods in Greece, simply due the attention accorded these time periods. Quite the opposite is true, however. What we find instead is an inconsistent record of the faunal remains. A handful of Classical and Hellenistic sites with analysis of faunal remains include the following: Corinth (The American School of Classical Studies at Athens), Lerna (Gejvall 1969), Tiryns (von den Driesch and Boessneck 1990), Asine (Moberg 1992), and Halieis (Jameson 1988) in the Peloponnese; Knossos on Crete (Reese, various publications), Kabirion (Jameson 1988) near Thebes, the Athenian Agora in Athens (The American School of Classical Studies at Athens); New Halos (Rienders and Prummel 2003) in Thessaly; Kalapodi (Stanzel 1991) in Phocis, and Kassope (Friedl 1984) in Epiros (*Table 6.6*).

Several factors may have contributed to the sparse faunal record in such a well-excavated time period, foremost among which must be the previously mentioned character of these sites - rich in other material artifacts, epigraphy and monumental architecture. Another contributing factor may well be time period during which many of the well-known sites, such as Corinth, were excavated. Many of these sites began excavations nearly one hundred years ago, during a time when faunal analyses were virtually unheard of and accorded little importance. Data that was kept on assemblages of this period tends to focus on animal bones in ritual and religious contexts, such as sacrificial offerings, and animal bones in the context of human burials.[41] Rather than criticize this unfortunate trend, we should view is as yet

40 SL = species list S = skeletal element frequency A = age determination
DR = detailed report on specific species M = measurements D = discussion/analysis
NISP = number of identified specimens MNI = minimum number individuals

41 See the ASCSA's numerous publications on the Athenian Agora.

another area in which to concentrate research, possibly gaining new insight into time periods exhausted by artifactual, architectural and textual studies.

SITE	*LOCATION*	*ANALYSES PEFORMED*[42]
Corinth	Peloponnese	SL
Lerna	Peloponnese	SL; M; S; A; D
Tiryns	Peloponnese	SL; M
Asine	Peloponnese	SL
Halieis	Peloponnese	SL
Knossos	Crete	SL
Kabirion	Thebes	SL
Athenian Agora	Athens	SL
New Halos	Thessaly	SL; S; D
Kalapodi	Phocis	SL
Kassope	Epiros	SL

Table 6.6 **Faunal analyses at Classical and Hellenistic sites (see text for references)**

Roman and Byzantine Sites:

Faunal analyses in the Roman and Byzantine periods have not fared much better than those in the Classical and Hellenistic periods. Taken together, there are a handful of sites for a span of over 1000 years. Most notable among these are the Athenian Agora and Ancient Corinth, excavations that originally had a Classical intent. The remaining sites should be viewed as very incomplete and unrepresentative of the wealth of material that is, or was, truly extant. They include the previously mentioned sites at Kitsos in Central Greece, Knossos on Crete; Lerna, Nichoria, Midea, Asine and Halieis in the Peloponnese; and Kabeirion sanctuary in Thebes.

Once again, it is not a lack of material that explains this, but rather a lack of attention. Traditionally, Roman and Byzantine material remains have been cast aside by scholars who have regarded their very existence a nuisance. In fact, many archaeologists have chosen to destroy this material in order to arrive at the more "interesting" remains underneath. It has often been dumped to the side, its presence noted and recorded, but not studied, quantified or retained. Nearly every Greek site contains a dump relegated to "Roman/Byzantine" garbage, and predictably, animal bones have also found their way there. The end result of such impoverished excavation methods has produced a very serious gap in our knowledge of

---

[42] SL = species list S = skeletal element frequency A = age determination
DR = detailed report on specific species M = measurements D = discussion/analysis
NISP = number of identified specimens MNI = minimum number individuals

faunal remains from these very seminal periods in Greece. They certainly warrant further study and attention, and could hold great potential to gain insight into often neglected and overlooked periods.

Reese (1994a.) has acknowledged the deplorable state of faunal analysis in the Roman and Byzantine eras of Greece, and advocates attention be paid to these later and understudied periods. Yet, I am unaware of any significant changes since the publication of his ideas, in the prevailing thought patterns of most archaeologists on Greek sites. It is unfortunate that those who are progressive in the field of zooarchaeology, such as Cosmopoulos, Halstead, Hamilakis, Reese and Gamble, still turn their attention elsewhere.

VII. Historical Methods in Greek Faunal Analysis

To truly understand the state of zooarchaeology in Greece, past and present, we must examine the questions and methodologies that have guided researchers in this field. Firstly, a critical evaluation of recovery methods, analyses and interpretations will be presented, followed by a discussion of the potential questions that faunal analyses in Greece may address in both prehistoric and historic periods.

As previously mentioned, recovery of faunal remains in Greek archaeology has been inconsistent. Prior to Gejvall's volume on the fauna of Lerna (Gejvall 1969), faunal reports were largely not considered a useful part of archaeological excavations. Lerna was one of the pioneering sites in this respect; however the methods employed in the recovery of these remains in 1958 have changed little.[43] Animal bone at Lerna, and at many sites in Greece, is recovered in the field by eye, usually by a hired worker. Dry sieving is often neglected, and wet sieving is even less common. Thus the amount and type of bone recovered is often biased toward larger, more easily recognizable fragments. There have been several actualistic studies, as well as re-examinations of excavated sites, in which the effects of wet sieving were examined, indicating a significant portion of overlooked bone in its absence (Lyman 1994; Payne 1995). However, this methodology shines in comparison to the many sites that simply did not collect animal bone.

Yet, we do have a few well-documented faunal collections from archaeological contexts in Greece, and it is beneficial to turn to the methods of analysis and guiding questions from these few assemblages. Traditionally, non-human osteological remains have

[43] Tsoungiza will be a noted exception to this generalization.

been given very cursory analysis, focusing on quantitative study and species lists. Even NISP (number of individual specimens present) and MNI/MNE (minimum number of individuals/minimum number of elements) analyses are relatively uncommon in the literature, usually replaced by a numerical breakdown of sheer numbers of bone, a list of extant species and some discussion of the exploitation of wild versus domestic animals. Research in addition to these simple analyses has focused on the role of animal husbandry in the overall economy and the management of herds and flocks (Halstead 1987; Pullen 1992). Research along these lines of inquiry has led to a flurry of discourse on the exploitation of primary and secondary animal products (Halstead 1996; Jones 1987; Pullen 1992; Sherratt 1981). In this vein, kill-off patterns of sheep and goat have been analyzed in order to determine whether individuals were primarily utilized for subsistence, or were used for their milk and wool, and only later eaten (Greenfield and Fowler 2003, 2005; Halstead 1981, 1986; Mancz 1989). Additionally, researchers are beginning to use faunal analyses in the reconstruction of past environments in order to approach questions surrounding climate change and plant husbandry (Bottema 1994; Curci and Tagliacozzo 2003; Hansen 1994; Valamoti 2004).

Today, many recognize the utility of faunal analyses, and as this chapter has shown, new methods are being utilized and applied in Greek assemblages. Yet, most research incorporating animal bones in Greece is stagnant. Many sites continue to neglect interpretive analyses. Zooarchaeology still occupies a liminal status on Greek projects, often it seems, being one of the first areas to fall prey to funding shortages. I would argue that this is due to the failure of Greek archaeologists to understand the potential that animal bones hold in addressing larger socio-economic questions – especially in smaller scale settlement contexts. It appears, therefore, that the present state of zooarchaeology in Greece is quite undeveloped. Research continues to focus on Bronze Age and earlier material, while hovering around the same questions concerning the use and effects of sheep and goat on the economy. Archaeologists working with Greek faunal material need to start asking new questions – of both their existing collections and of their current and future excavations.

VIII. Conclusion

Greek archaeology must regularly examine the role of all animals in archaeological assemblages. It must begin using the faunal remains to approach larger theoretical issues, such as socio-economic complexity. This study is an effort in just this. Thus far, much of what we know of early Greek economies in regard to livestock survives from analysis of Linear B tablets of Mycenaean date (Halstead 1987, 2003). While useful, the tablets paint a picture focused on the palatial economy, and the future of zooarchaeology in Greece will benefit most by using faunal remains to address intra-site socio-economic issues at a local, small-scale level. While discussion arises occasionally, the socio-economic potential of zooarchaeology in Greece remains a largely unexplored field.

As shown, many of the larger archaeological projects in Greece during the last century have focused on the more spectacular, visually pleasing elements of ancient Greek society. To this end, attention was directed toward excavating the Athenian Agora, Ancient Corinth, Lerna, Pylos, Knossos, Mycenae, Tiryns, Akrotiri, and any site that held promise for museum worthy antiquities and visitors. Attention, for the most part, has not been trained on smaller-scale settlements – with some exception given to the Neolithic period. It is in the arena of small village life that faunal analyses can provide much-needed new insight. Now armed with an understanding of the state of faunal analysis in Greece, the following chapter is an effort in changing the static state of Greek zooarchaeology. It presents the faunal data from the Early Helladic assemblages at Helike, Lerna, Tsoungiza and Tiryns on which the ensuing analyses of socio-economic complexity are based.

CHAPTER 7

EARLY HELLADIC FAUNAL ASSEMBLAGES AT HELIKE, TSOUNGIZA, LERNA AND TIRYNS

Inter-site faunal analyses are one method with the potential to ascertain whether evidence for social complexity exists in the raising of livestock, and address the hypothesis that faunal remains address economic aspects of social complexity in Early Bronze Age Greece. This chapter systematically examines the faunal assemblages from Early Helladic Helike, Lerna, Tsoungiza and Tiryns. It provides a brief background for each site, followed by a detailed description and analysis of the faunal data. Chapter 8 will carry out an in-depth analysis and discussion of pigs in particular, and will then interrogate the data within the framework of socio-economic complexity.

I. Helike

Physical and Geographical Location:

Helike is an Early Helladic (EH) settlement of EH II-IIIA (2600-2300 B.C.) date located on the south west shore of the Gulf of Corinth in the northern Peloponnese (*Figure 2.1*). It lies in a coastal plain, about 1 km inland in the modern village of Rizomylos, in the province of Achaia. It is bordered to the northeast by the Gulf of Corinth, to the south by rocky foothills, to the west by the Selinous River and the east by the Kerynites River. The site itself is located in flat agricultural fields, among modern olive orchards and grape vineyards. Architectural remains of Early Helladic date were initially uncovered at the site in 2001. These remains lie below the standing water table, approximately three to five meters below the ground surface.

Analysis of sediment samples taken from within the Early Helladic trenches yielded evidence for both marine and lagoonal microfauna. This suggests that the ruins had been submerged in sea water for some time. Early Bronze Age (EBA) Helike was originally a coastal settlement. However the area was, and continues to be, highly seismic, resulting in significant geological change in the location of the modern shoreline. The site itself lays on the Helike Fault, a major fault line that runs laterally along the length of the Gulf of Corinth. Seismic activity probably played a significant role in the history of Early Bronze Age Helike. Excavation of one of the foundation walls from a structure within the Early Helladic settlement is abruptly offset by what appears to have been seismic discontinuity. This suggests that the Early Helladic settlement may have been destroyed by an earthquake and then submerged as the land mass slumped.

The Early Helladic Settlement:

Excavation of the Early Bronze Age settlement at Helike began in 2001. Collaboration between the Patras Ephoreia, the University of Patras, New York University and the Helike Foundation, resulted in field seasons in 2001, 2003, 2004, and 2005. These seasons explored Early Helladic contexts in an area measuring approximately 40 m x 5 m. This area consists of four contiguous trenches, each roughly 10 m x 5 m in size. Cultural material commences at a depth of about three meters below the ground surface, and this has consisted almost exclusively of pottery sherds. Complete vessels have yet to be recovered at this depth.

Conditions at Helike make excavation difficult. The ground water table for the area begins at a depth ranging from 3.2 m to 4 m. The later in the summer that excavation can occur, the lower the water table due to very dry conditions in June, July and August. Even so, conditions in the trenches are waterlogged, creating the need to continually pump water out of the trench on a regular basis. This means that excavation never occurs in dry conditions, even during seasonally dry periods, and is still carried out in wet mud. More will be said on this during the discussion of the fauna.

Early Helladic architectural remains begin to appear at a depth of about four meters. These consist of stone foundation walls, stone "platforms" and pebble "paved" alleyways or cobbled streets. At this depth, complete ceramic vessels have been found. These vessels include large pithoi (storage jars), cooking pots, tankards, jugs and a variety of eating and drinking vessels. Preliminary analysis of the pottery in the field indicates that the majority were handmade, medium-coarse ware vessels. Occasional wheel-made, fine ware fragments were also unearthed. Many of these fine wares have a dark on light design, closely mirroring pottery decoration found on the island of Aigina (north east of the Peloponnese).

Though the architectural remains are still under study, their preliminary analysis provides insight into the nature of the settlement at Helike. Remains of stone foundation walls imply a settlement consisting of several adjoining rooms or buildings. Returns on some of the walls suggest the presence of at least two separate structures, each consisting of several rooms. While the exact function of each of the rooms and/or structures is not yet fully understood, they do appear to be the remains of domestic housing. The presence of a hearth, a storage/garbage pit (bothros), stone platform, burned pottery and burned animal bone with cut marks, suggests the remains of cooking and food preparation. Additionally, fragments of decayed mud brick are found throughout the trenches. While the foundations of the structures were of stone, the walls would have been made out of mud brick. No roof tiles have been found, suggesting thatched or wooden roofs. The walls themselves vary in thickness, but are nearly all approximately 50 cm in width, with preserved heights ranging from 20 to 50 cm. This suggests that the structures would have been single-storied.

At other sites throughout the Early Helladic II-III period in the Peloponnese, structures known as corridor houses have been found. Portions of the excavated foundation walls at Helike have been hypothesized to represent a building of similar form (D. Katsonopoulou, personal communication). However, excavation has not been able to uncover the full footprint of any one structure in its entirety at Helike. Wall returns have been found indicating the presence of several structures, but each of the structures extends into the baulk of the trench walls. Private ownership of the adjacent agricultural fields prevents excavation in these directions at this stage.

Geophysics has been used extensively at Helike. Geophysical tests of the seismic reflection method at the Early Bronze Age site produced images of extensive probable buried structures having the same depth (four to five meters) and orientation as the walls from the adjacent, excavated trenches. Until such time when excavation can be extended into the surrounding area, definitive conclusions as to the physical reconstruction of the Early Helladic structures can not be made. Geophysical testing strongly suggests the presence of additional architectural remains, but at present, analysis of the material remains offers the soundest method for understanding the nature of the settlement. These remains consist mainly of pottery and animal bone, with additional stone tools and lithic debotage. Analysis of spatial patterning of the material remains and their location in relation to foundation walls will provide information as to the potential use and function of the individual "rooms" within these larger structures.

One final note about the architectural remains at Helike need be made, and that is the existence of several building and/or occupational phases. During the 2004 and 2005 field seasons the author was the field site supervisor. Excavation yielded evidence for more than one phase of building, based upon the relative depths of the foundation walls. The bases of many of the excavated walls varied in depth, ranging from a resting depth of four meters from the ground surface, to over six meters below the ground surface. Many do not run parallel or join other walls, and several are built over the top of earlier walls. Some robbing of stone appears to have taken place during construction of later settlement/building phases, making it difficult to match walls or structures of earlier and later date. While it is certain that several phases of building took place, the chronological span of time represented by these phases is not known. Building and rebuilding could have occurred over one year, 20 or 200. At this point in time, further excavation is required to address this issue.

In sum, all lines of evidence uncovered so far at Helike suggest the existence of an Early Helladic settlement. Several intact pithoi and smaller storage vessels suggest the presence of a "store room", while the presence of a hearth, bothros and stone platform with associated burned pottery and animal bone suggests the remains of domestic food preparation. Due to limitations on the area available for excavation, several unanswered questions exist surrounding the exact size and nature of the site, building types, and even its relative importance and relationship to other sites in the area. Faunal analysis offers one of the best ways in which some of these questions can be addressed given these current limitations.

The Faunal Remains:

Excavations from the 2001-2005 field seasons at Early Bronze Age Helike yielded 382 identified specimens (NISP), from a total faunal assemblage of 729 specimens. 52 percent of the assemblage was identified by the author over the course of two field seasons in 2004 and 2005. This assemblage is the result of an excavated area of 5 m x 40 m,[44] and based on pottery seriation, has been dated to a single chronological period, Early Helladic II/IIIA. While the assemblage is the smallest of those under analysis in this dissertation (*Figure 7.1*), it yields valuable information about the nature of the settlement at Helike. It provides a snapshot of the animal economy in the excavated area of the site.

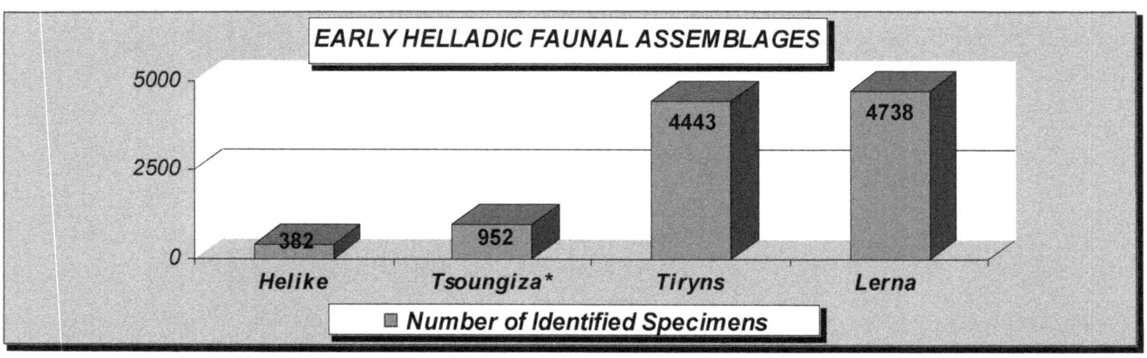

Figure 7.1 **Early Helladic Faunal Assemblages (NISP)** Lerna III/IV (Gejvall 1969:10); *Tsoungiza data is quantified as MinAU (Halstead, in press); Tiryns (von den Driesch and Boessneck 1990); Helike (Fillios 2006).

[44] Four 10 m x 5 m trenches - H22, H38, H43, H51

Excavation continues at Helike. The data under analysis here provides a starting point. As the size of the excavated area grows, and the size of the faunal assemblage increases, a more complete image will emerge. Concerns about sample size at Helike are still present, however. This study will address these concerns by looking at species representation in proportions. That is, it takes assemblages from similar sites and similar contexts and compares the relative proportions of species and skeletal elements. The results of this process will be discussed in the following chapter. The issue of sample size is highlighted, however, because a small assemblage has the potential to be influenced by many biases, and may not therefore be representative of the site and/or assemblage as a whole. Without further excavation, there is little that can be done about the relatively small faunal collection at Early Helladic Helike, and so the best way in which to mitigate any resulting problems is through thorough analyses that account for these potential problems.

The faunal assemblage analyzed covers the 2001, 2003, 2004 and 2005 seasons of excavation at Helike. The 2001 season at Helike was a test trench on a smaller scale than ensuing field seasons, and thus yielded smaller quantities of faunal remains. The 2003, 2004 and 2005 excavations yielded higher amounts of bone due to more extensive exploration.

The animal bone consists of three main species: *Bos* (cattle), *Sus* (pig) and *Ovicaprines* (sheep/goat). *Cervid* (deer), *Equid (possible donkey)* and possible *Canid* (dog) and human remains were also identified. Pig accounts for the majority of the identified bone, followed by sheep/goat and cattle.

Methodology:

The faunal material examined in this chapter is the result of four field seasons of excavation, 2001, 2003, 2004 and 2005. My dual roles as site field supervisor and faunal analyst during the 2004-5 seasons allows me to directly account for excavation methods during these two years. During the 2001 and 2003 seasons, the recovery of bone was limited to hand collection in the trench by workmen. In the two seasons I was present, while hand collection of bone continued, I initiated the further method of wet sieving the excavated soil. Approximately 10 percent of the soil excavated was wet-sieved with a 10 mm mesh. Wet-sieving was the only option for screening as the soil is water-logged. In selected contexts such as hearths, bothroi or particularly dense concentrations of bone, the entire deposit was wet-sieved.

Bone was bagged in the field by context, stratigraphic level, and area, such as Room II. Note of exact provenience was also made, when possible (such as below platform one, next to pithos two, or adjacent to wall one). In addition, material recovered from wet-sieving was distinguished from that recovered by hand in the trench. Since Early Bronze Age Helike is a single-phase site from a chronological perspective, further sorting in the field was not necessary. Bone was transported back to the lab in the village of Helike, where it was cleaned by me with water. Further cleaning was carried out with acetone when needed, and in the case of poor preservation, bone was dry-brushed. After cleaning, the bone was rebagged and underwent identification and recording. This was done in two stages:

Stage 1: The contents of each bone bag were sorted into general anatomical groups. These groups consisted of four main categories:

- skull (cranium, mandible, loose teeth, horn/antler),
- trunk (ribs, pelvis, vertebrae, sacrum),
- limb (scapula, humerus, radius, ulna, femur, tibia, fibula) and
- foot (metacarpals, metatarsals, phalanges).

At this stage, note was made of any evidence for the presence of whole or partial individuals, for groups of fragments skewed toward specific body parts, for fresh breaks that occurred in the field, and for the presence of 'new' and perhaps intrusive bone.

Stage 2: Specimens potentially identifiable to specific anatomical parts (i.e. femur) and to taxon (i.e. pig, cow) were separated for detailed examination and recording. Anatomical parts selected for identification were: horncore, antler, skull, maxilla, mandible, molar, premolar, canine, incisor, vertebra (cervical, thoracic, lumbar – identification permitting), scapula, rib, pelvis, sacrum, humerus, radius, ulna, metacarpal, phalange, femur, tibia, fibula, astragalus, calcaneus, metatarsal, metapodial, and long bone shaft. These elements were further described by part represented: proximal, medial, distal, anterior, and posterior. Distinct anatomical features were also recorded, such as the presence of a discernible linea aspera, a scapular glenoid fossa or pelvic obturator foramen. Phalanges were not always identifiable to number, nor were fore-limb and hind-limb phalanges distinguished. The preceding anatomical parts were utilized for the information they contain on taxon/species identification, and for informative parameters such as age and sex. When applicable, measurements were made on selected specimens following the parameters and procedures set out by von den Driesch (1976).

The 382 identified specimens from Early Bronze Age Helike represent only those bones identified to both anatomical part and taxon. After this preliminary identification, any additional variables were noted and recorded where appropriate. These variables include: side of body, presence/absence of proximal and distal ends, state of epiphyseal fusion, dental development/wear, sex, pathological traces, fragmentation, evidence of rodent/carnivore gnawing, burning, cut marks, root etching, other taphonomic factors, and metrical data. All identified bone specimens at Helike were given a catalogue number, labeled, and placed into an Excel database. Any specimen containing distinguishing features (i.e. pathology, traces of butchery) was photographed and linked to the database entry.

The faunal assemblage was examined and analyzed in the Helike Project's laboratory/storerooms in the village of Helike, Achaia. Identification was aided with the use of a small modern reference collection, procured by the author from the surrounding area, and standard published manuals. A 10X magnification hand loop was used for closer examination of bone surface modification, and measurements were taken with 150 mm digital calipers.

The animal bones recovered from Helike were generally in a poor state of preservation, and their fragile condition was further compounded by the water-logged conditions in the trench. These factors have a direct impact on the faunal analysis carried out. Analysis indicates that a great majority of the bone was broken after deposition. Fresh breaks were common due to lack of visibility in the mud, and poor preservation often caused smaller, less dense bone to crumble upon retrieval. This created a very fragmentary assemblage in which small bones were often rendered unidentifiable. Breakage in the field also affects the final sample size and percentage of identified versus unidentified bone. The number of bone fragments appears to be greater than would originally have been the case, had better field conditions prevailed. In addition, water-logged conditions may have created an identification bias toward the more robust, denser bones. While some information was undoubtedly lost due to factors beyond our control, conducting detailed analyses will seek to minimize this impact on the interpretation of the assemblage.

Age at Death:

When possible, age at death of the animals at Helike was estimated on the state of tooth eruption, tooth wear, and epiphyseal fusion (development stage of the long bones). Established tooth wear stages and eruption data are available for most of the major domesticates (Halstead 1985; Grant 1982; Wilson et al. 1982). Dental ageing for cattle follows Grigson (1982), for sheep/goat (Deniz and Payne 1982; Payne 1973) and for pig (Bull and Payne 1982). Ageing of post-cranial material follows (Silver 1969; Wilson et al. 1982). Metrical data (von den Driesch 1976) plays a subsidiary role in age determination, and also in determination of sex. In addition, the assemblage as a whole was analyzed (subdivided by species) to address the following demographic parameters: mortality age profiles, seasonality, culling patterns, and sex ratios.

Taphonomic factors (human and natural):

Each bone specimen was analyzed for evidence of several pre- and post-depositional agents. Identification and analysis of bone surface modification by carnivores, rodents and humans, was severely affected by root etching and heavy erosion of the surfaces of a large portion of the assemblage. Roots from the lagooning over of the site have acted on much of the bone, eroding their surfaces. Approximately 72 percent of the assemblage exhibits evidence of root etching and/or weathering.

The ph of both the water and the mud in trench H51 was tested, and found to be neutral (seven). This ruled out current acidic conditions, but not necessarily those of the past. A soil ph of seven (neutral) could potentially be a "false" reading resulting from eroding alkaline limestone sediment (limestone cliffs just above the site) seeping into the water table. This sediment, mixed with the acidic environment in the trench itself (produced by decaying roots and organic matter), may result in a "false" basic reading. In cases where root etching was not a large factor, specimens were examined for evidence of rodent/carnivore gnawing (rarely noted), burning, and cut marks resulting from butchery.

Cut marks were discernible on several specimens, and in each case, a cast/impression of the mark itself was taken. The bone surface was cleaned first with alcohol, followed by acetone, to ensure the clearest impression would result. Impressions were taken with liquid dental silicone which was forced into the mark with a special syringe. The liquid silicone hardens within minutes, and when peeled off, results in a raised, negative impression of the cut mark itself. A positive image of the mark can later be made with modeling clay for use in a scanning electron microscope (SEM). The cut mark impressions taken can be analyzed with the aim of ascertaining the type of tool (stone or metal) that inflicted the cut. After silicone impressions were completed, photographs of each specimen were taken and anatomical cut mark placement was noted. While a detailed analysis of cut marks at Helike is planned, only general discussion of cut mark distributions are employed in this study in regard to butchery patterns.

Burning on bone surfaces was noted, but analyzed cautiously and in reference to provenience of the specimen in question. It is hypothesized that Early Helladic Helike was destroyed by the combination of an earthquake and fire; thus, burning on bone may have resulted from site destruction, and not food preparation.

Quantification:

At Helike, several quantitative methods were applied to the assemblage:

- NISP (number of identified specimens/taxon),
- MNI (minimum number of individuals), and
- MNE (minimal number of elements)

These methods were calculated to facilitate comparison of the Helike assemblage with others (*Figure 7.2*).[45] Analysis of published Greek faunal reports shows that such quantitative analyses are not standardized. Some sites calculate MNI and others provide simple specimen counts (NISP). In addition to the lack of quantitative standardization, different analytical methods may hold more or less merit depending upon questions posed of an assemblage. Therefore, to benefit future research, several types of quantitative analyses on the Helike fauna were performed.

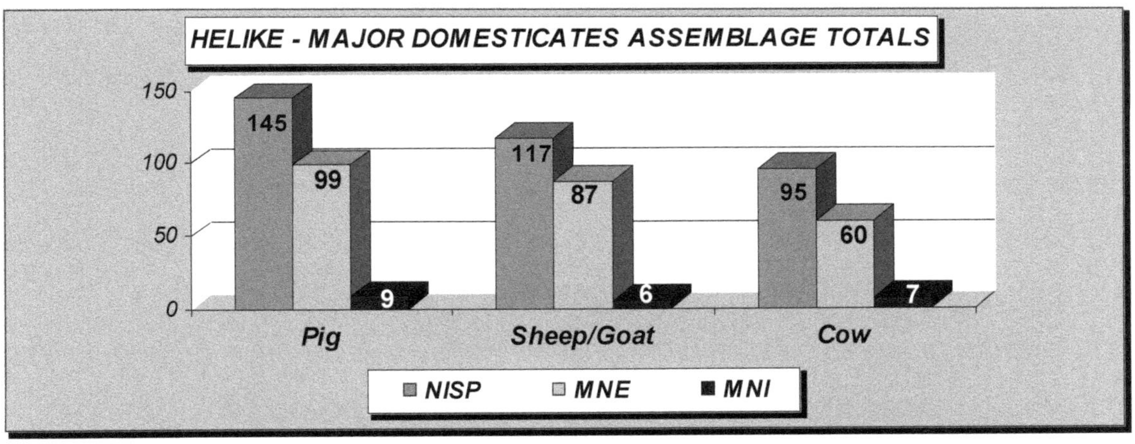

Figure 7.2 **Helike faunal data: Assemblage totals for major domesticates NISP:MNE:MNI (see appendix Tables 1-4)**

MNI was derived at Helike with several variables taken into account. Sexing and ageing data were used in estimates when available, as was side of the body. In addition, if variability in size was present for two of the same body parts upon visual inspection, i.e. a distal and proximal humerus that varied drastically in size suggesting two different individuals, the MNI was calculated as two, rather than one. MNI frequencies were

[45] Calculations were based on methods described in Lyman (1994) *Vertebrate Taphonomy*. Cambridge.

calculated according to left, right and unsided. The total MNI count per element is then given as MNI per skeletal portion, and it is maximized.

MNE calculations were derived from MNI results, and are presented as MNE comp, meaning shaft fragments and epiphyses were taken together to arrive a final counts. In both MNI and MNE calculations, spatial distribution played a vital role in quantification. For example, if a left distal humerus and left proximal humerus identified as pig were present, but were found in different contexts, the MNI was calculated as two. NISP, MNI and MNE calculations are after Lyman (1994). MAU was not calculated on the faunal assemblage from Helike (Binford1981; Lyman 1994). This is because it is not used in the assemblages under comparison, and thus would not facilitate inter site comparisons, nor add significantly to the final interpretation of Helike. While at Tsoungiza MinAU is used as the main method of quantification, it is akin to MNE, and thus MNE calculations can be used in comparisons. This will be addressed in greater detail in the ensuing pages. Due to the lack of standardization among published faunal material, note will be made of the quantification method employed in all inter site comparisons.

Size and Overall Composition of the Early Helladic Assemblage:

The faunal material from Early Helladic Helike is comprised of 729 total specimens, of which 382 were identified, or approximately 52 percent. It has been suggested in the literature (van der Veen and Fieller 1982) that a sample size of around 400 is needed for an accurate (within five percent) and reliable estimate of taxonomic and/or anatomical composition. The small sample size under study here has its limitations in this respect. Further, a small portion of the sample offer information on butchery (eight percent), sex, age, and possible food preparation (13 percent burned).

Three main species account for the majority of the animals present in the assemblage. Pig bones account for 37.9 percent of the identified animal remains (NISP). They are followed numerically by sheep/goat (30.6 percent) and cattle (24.9 percent) (*Figure 7.3*). Due to the often fragmentary nature of the assemblage, metrical distinction between sheep and goat was rarely possible. They are here treated together, and when conditions permitted, their discernment is noted. In addition, metrical distinctions between wild and domestic pig are not attempted, nor is further refinement of cattle.

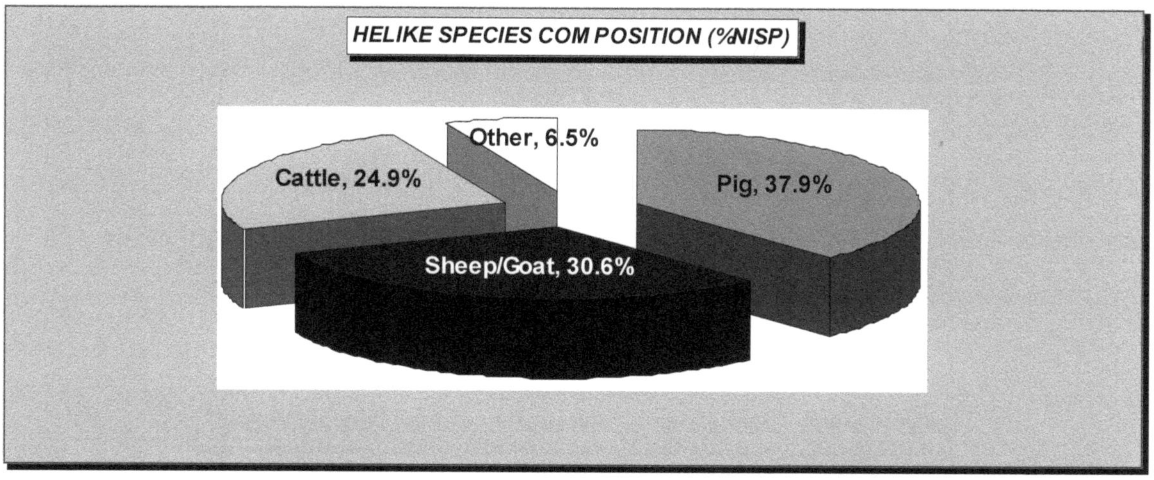

Figure 7.3 **(see appendix Table 1)**

Body part representation was also examined for the major species at Helike. When subdivided into four main body part groups, the data for each of the domesticates is similar. All skeletal elements were divided into:

- cranial (horn cores, skull, mandible),
- axial (ribs, pelvis, vertebrae, sacrum),
- appendicular (scapula, humerus, radius, ulna, femur, tibia) and
- foot (metacarpals, metatarsals, astragalus, calcaneus, phalanges)

This division was done to facilitate analyses of body part representation on a broad scale. Figure 7.4 illustrates that for each species, all major body parts are present. This suggests that carcass processing and animal consumption took place on site.

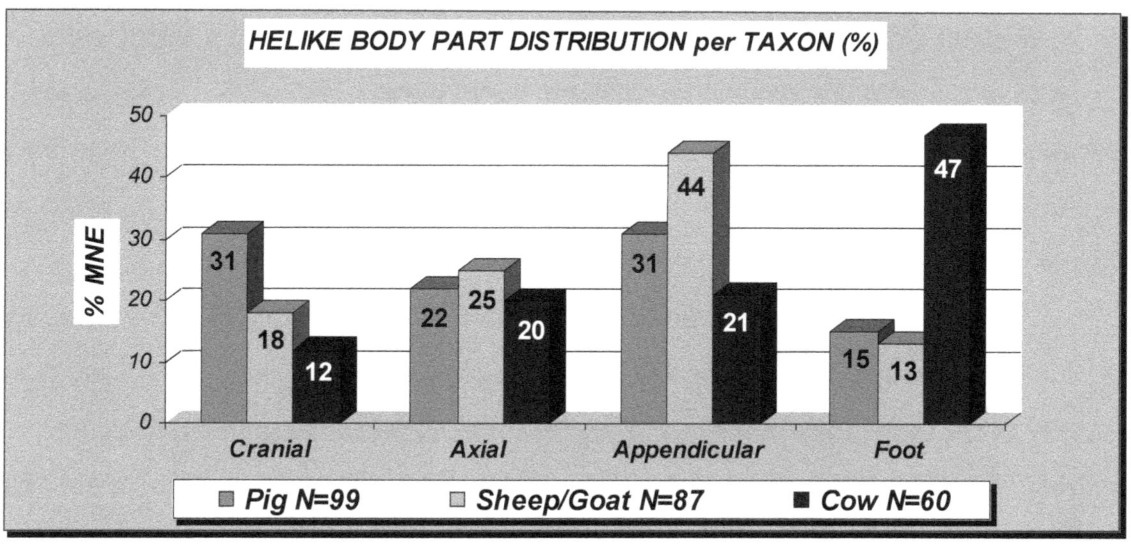

Figure 7.4 **(see appendix Table 6)**

Sus:

Pig bones account for nearly 38 percent of the identified species when viewed based on NISP frequencies (*Figure 7.3*). Examination of both MNI and MNE calculations indicates that they continue to form the same percentage of the total assemblage. NISP/MNE ratio examination shows the two methods to be fairly similar in result, indicating that the assemblage is not as fragmentary as originally thought (*Figure 7.5*).

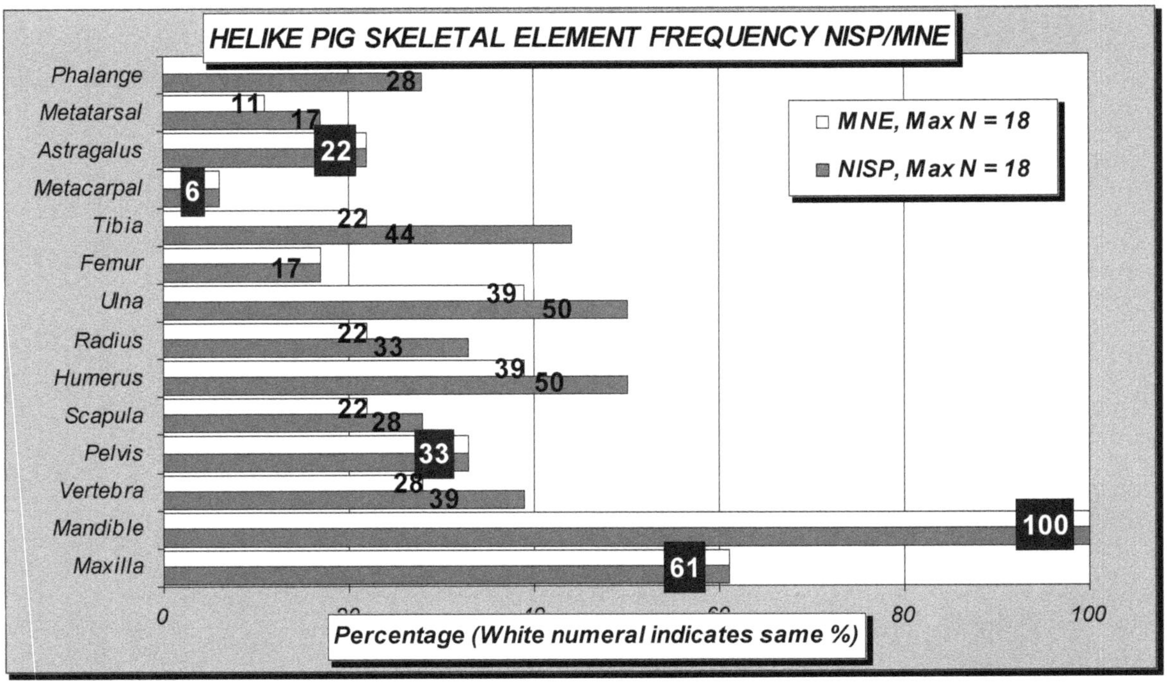

Figure 7.5 **Helike pig skeletal element frequency NISP%/MNE % (see appendix Tables 1 & 4)**

Skeletal element frequency of pig bones shows a distribution of all body parts (*Figure 7.5*). Cranial, axial and appendicular elements are all present, indicating that the animals are probably being raised and slaughtered on site. This has wider implications in trade and economic exchange, as it suggests that Helike was most likely not obtaining pigs from other villages, but instead practicing localized animal husbandry.

Pig attrition and mortality profiles offer information regarding Helike's subsistence and animal economy. Mortality profiles based on tooth eruption/wear data indicate that the majority of the individuals in the assemblage were older – over 24 months of age (60 percent). Data derived from teeth also show a handful of individuals in both the 6-12 month and 12-24 month age ranges (20 percent each) (*Figure 7.6*). This data offers several interpretive possibilities. It could be that the prime age for slaughter was over 24 months at Helike when breeding and rearing factors were taken into consideration. A high infant mortality rate could also have been a factor. Another possibility, however, is suggested by data based on profiles generated from epiphyseal fusion.

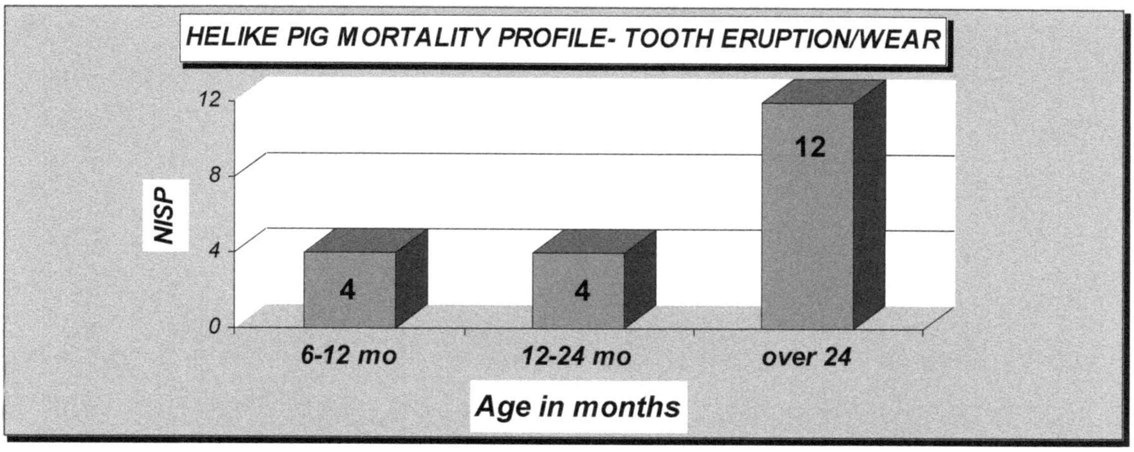

Figure 7.6 **(see appendix Table 10)**

Mortality profiles based on month of epiphyseal fusion indicate that the bones and teeth come from a majority of individuals over 24 months of age (nearly 90 percent) (*Figure7.7*). Prime aged individuals – 12-24 months of age are absent using this method, and a few younger individuals are present. This mortality profile suggests local animal husbandry with possible provisioning of other sites. Since the prime-aged individuals are absent, it is possible they were raised at Helike and sent elsewhere – either through trade or reciprocity. If this is the case, it suggests that Helike was involved in inter site economic relationships with neighboring settlements.

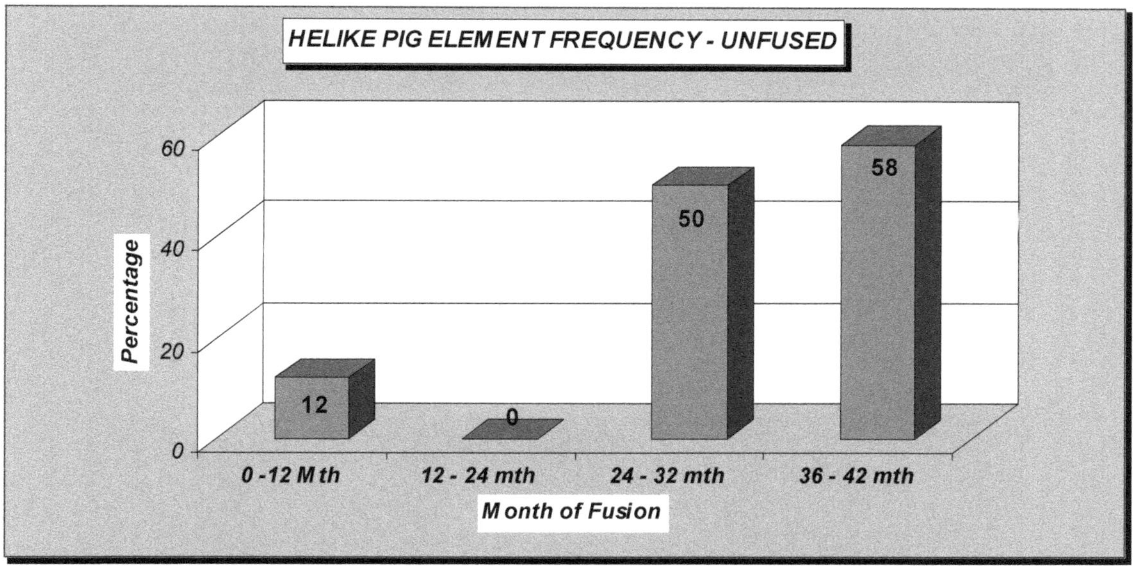

Figure 7.7 **Helike *sus* mortality profile based on epiphyseal fusion data %MNE (see appendix Figures 8-9)**

Hypotheses based on data derived from both ageing methods could also be subject to yet another interpretation. Due to the small portion of the site excavated, it is difficult to ascertain with certainty whether the fauna provide a snapshot of just one portion of the site. Does this represent subsistence remains from a few households or in fact reflect subsistence and economic variables from the site as a whole? If the data do represent just a few households, the conclusion that prime aged pigs were being sent away as part of a trade/exchange network becomes tenuous. It could be that we are seeing the households/individuals who are raising the pigs for the *site*, and provisioning other households within Helike. Perhaps there is intra-site specialization and economic production. Unfortunately, intra-site questions can not be addressed until a larger portion of the site has been excavated.

Analysis of pig remains yields limited evidence on possible butchery patterns. This evidence can be added to that derived from mortality profiles, to better inform issues related to pig management. Although only nine percent of the total pig NISP bear traces of butchery, and seven and a half percent of the total pig MNE, when MNI is calculated for these specimens and then viewed as a percentage of the total pig MNI, this percentage increases to 100 percent. The relatively small number of bones with cut marks should be viewed critically, however. The majority of the faunal assemblage at Helike was subjected to heavy root etching which rendered most of the bone surfaces un-analyzable. Thus it is highly likely that more bones possessed traces of butchery that have now been lost.

Further, analysis of the skeletal element frequency against cut marks indicates at least primary butchery on site (*Figure 7.8*).

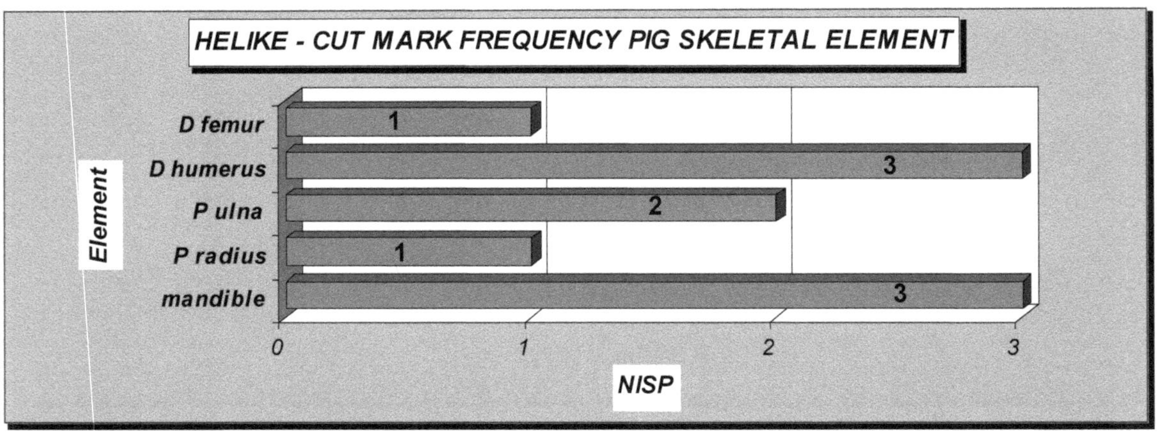

Figure 7.8 **(see appendix Table 14)**

Cut marks frequently occur at articulation points – thus marks on a distal femur could have occurred from disarticulation of the femur from the tibia. Mandibular cut marks occur on the hinge (separation from skull) and the interior surface (tongue). When viewed against the data derived from mortality profiles, further weight is given toward local rearing and animal husbandry of pigs.

Evidence of burning is also present on the pig assemblage, accounting for eight percent of the total NISP. Thirty-eight percent of the burned specimens also have evidence of butchery. Several were found associated with burned pottery and a possible hearth. Unfortunately, there is evidence of widespread burning at the site, and so only those fragments that were in direct association with a food preparation and butchery can certainly be taken as the remains of cooking.

Bos:

Cattle comprise nearly 25 percent of the identified species at Helike (*Figure 7.3,* p.113) based on NISP. MNI calculations show similar abundances in the assemblage. As with pigs, all body parts are fairly evenly represented, with cranial and post-cranial material suggesting local butchery and consumption in the settlement.

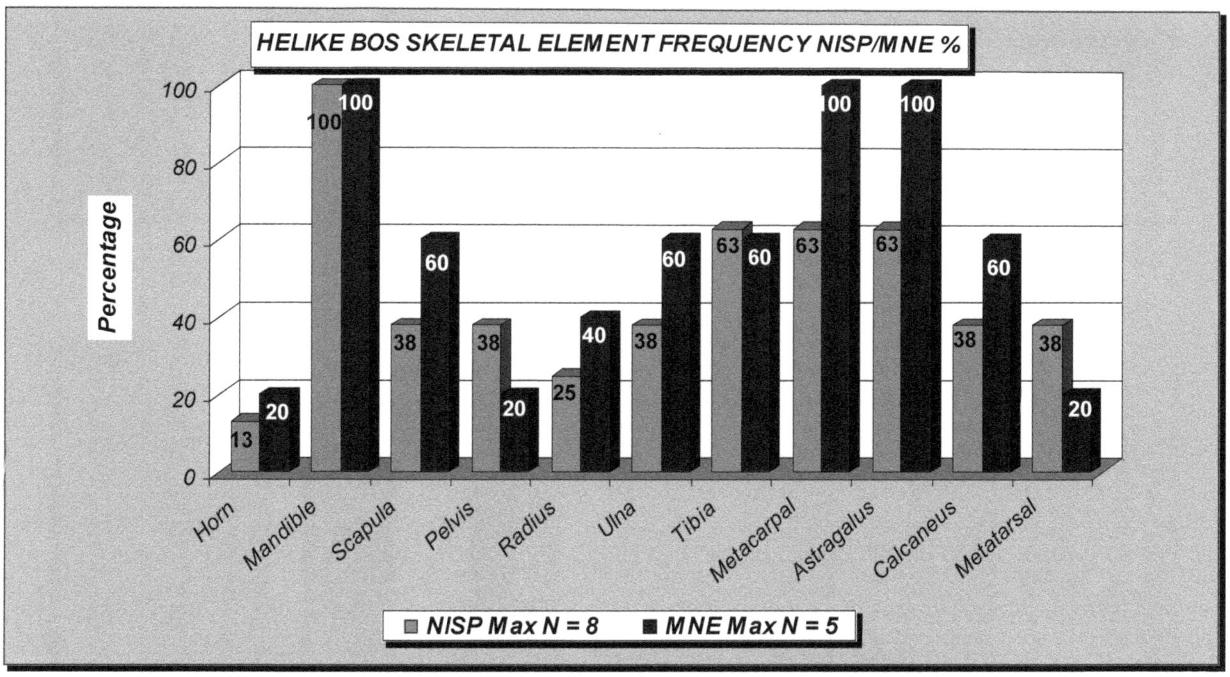

Figure 7.9 **Helike *bos* skeletal element frequency %NISP/%MNE (see appendix Tables 1-2)**

While examination of NISP/MNE ratios suggests a more fragmentary assemblage than pigs (*Figure 7.9*), the assemblage as a whole does not appear to have been mediated by carnivore activity. The fragmentary nature of the cattle bones could be due to human activity, such as marrow extraction. Analysis of breakage patterns that might provide insight into this question have yet to carried out on the assemblage.

Cattle attrition profiles and mortality data indicate that the majority of the individuals in the assemblage were over 24 months of age. Similar results were obtained from both dental data and epiphyseal fusion (*Figures 7.10* and *7.11*). The existence of a small number of younger individuals viewed alongside this data suggests that cattle were probably reared on or in the vicinity of the site. Interpretation of mortality data supports the conclusion offered by skeletal element frequencies. That is, local animal husbandry was responsible for the existence of cattle at Helike. Data from both cattle and pigs then suggests that with respect to these two species, Helike may have a self-contained animal production "economy".

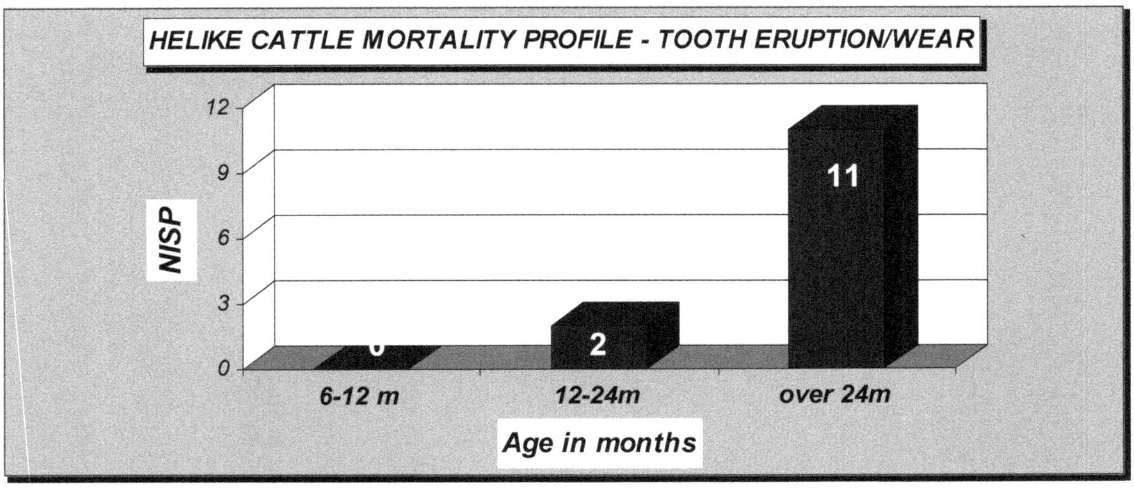

Figure 7.10 **(see appendix Table 10)**

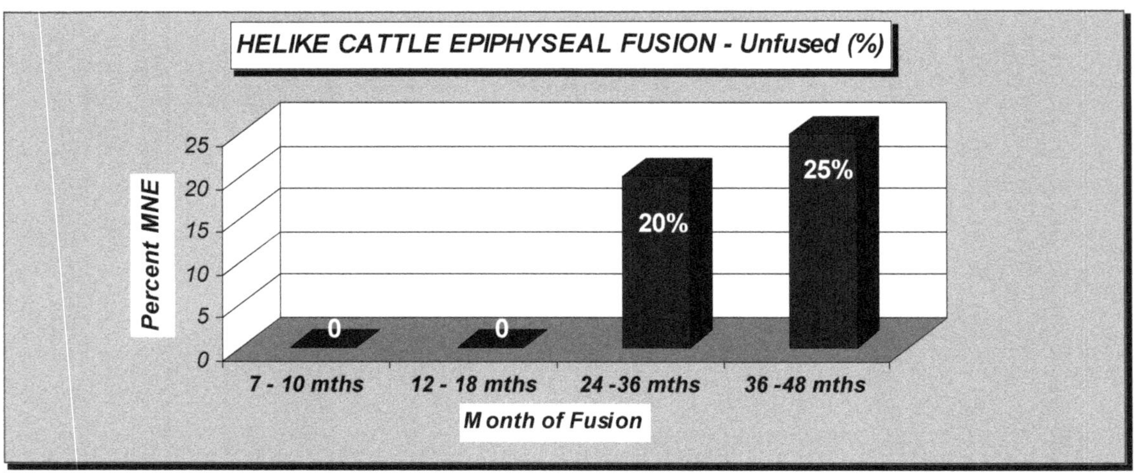

Figure 7.11 **(see appendix Figures 10-11)**

At Helike a small percentage of the cattle bones provide evidence for butchery practices (10.5 percent) if calculated by NISP. When viewed as a percentage of the total MNE, this percentage increases to 20 percent. As with pigs, the cut marks appear to result from primary butchery (disarticulation), with a few resulting from possible food preparation or secondary butchery. Analysis of skeletal element frequency against cut mark frequency does not indicate any discernible patterns (*Figure 7.12*).

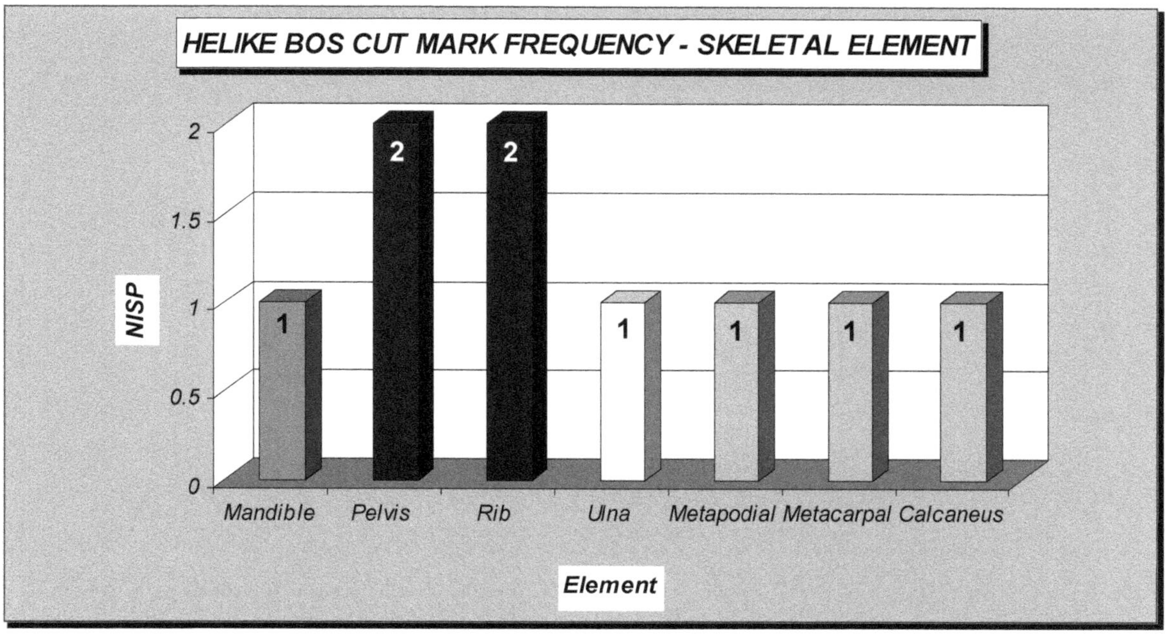

Figure 7.12 **(see appendix Table 14)**

Cattle bones also show evidence of burning, but as with pigs, it can not be determined for the majority whether this burning results from food preparation or site destruction. Two rib fragments bearing cut marks are also burned.

Ovicaprid:

At Helike sheep/goat fragments account for 30.6 percent of the total identified species as calculated using NISP (*Figure 7.3,* p. 113). This percentage differs when calculated by MNE and MNI. MNI calculations suggest a lower frequency of sheep and goat as compared to cattle. This suggests that sheep/goat bones are more fragmented than the other two major species. When MNI counts are used, sheep/goat account for 27 percent of the identified species, while cattle account for 32 percent. However, if MNE is used, the relative species ratio is closer to the NISP percentages – that is 35 percent *ovicaprid*, 25 percent *bos*, and 40 percent *sus*. As with pigs and cattle, sheep/goat remains are represented by nearly all body parts (*Figure 7.13*), once again suggesting on-site mortality.

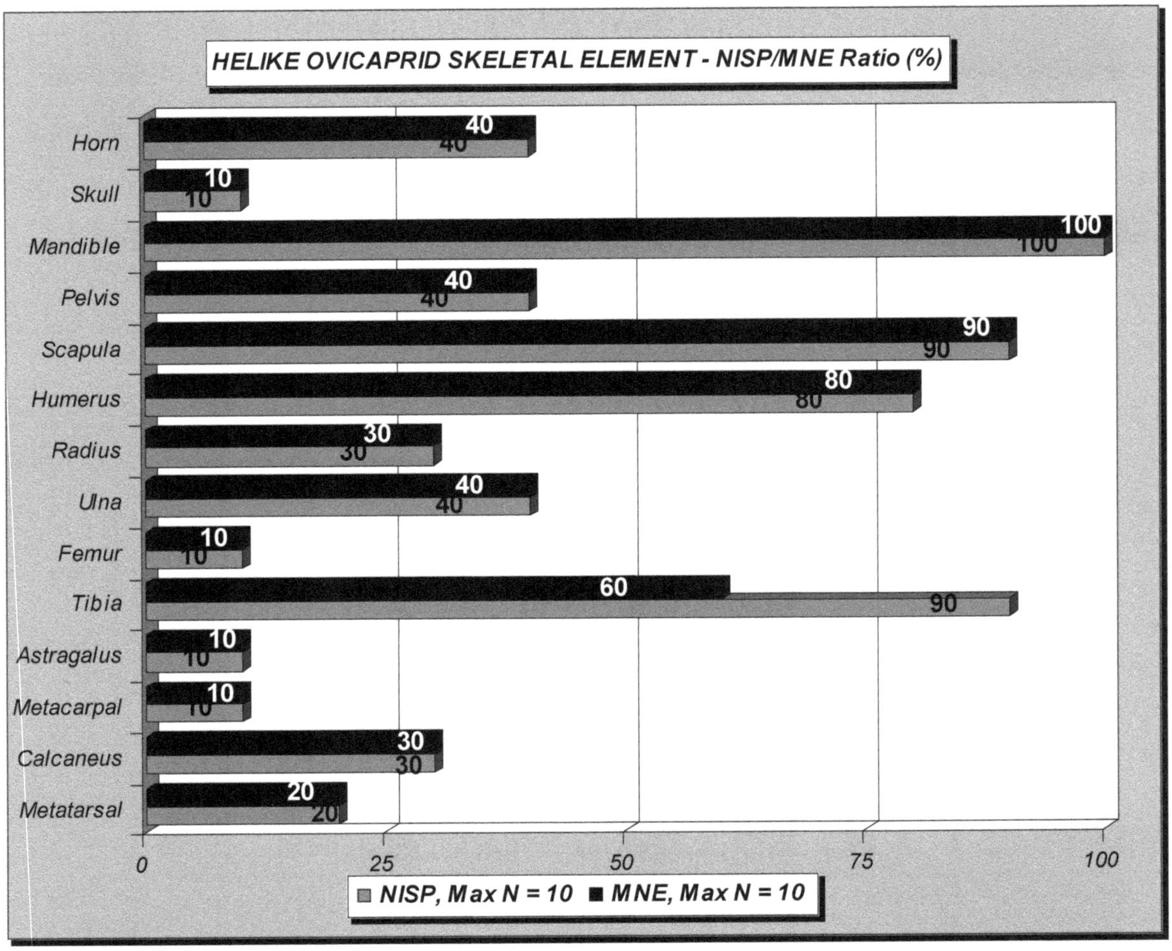

Figure 7.13 **Helike *ovicaprid* skeletal element frequency %NISP/%MNE (see appendix Tables 1& 3)**

Data from attrition and mortality profiles suggests shows a similar age distribution to cattle, with all ages present. Profiles based on dental data suggest that nearly 50 percent of the individuals were over two years of age (*Figure 7.14*). Data derived from epiphyseal fusion corroborate this finding (*Figure7.15*). The existence of all age ranges, coupled with a majority of older individuals could be open to various interpretations. It is possible that some individuals were kept as breading stock and others used in the production of secondary products (milk, wool). It is also possible that the bones from younger individuals did not survive due to lower density and higher fragility. However, the profiles once again suggest that sheep/goat were being reared and consumed locally, as with pigs and cattle. Data derived from ovicaprids then lends further support to the suggestion that Helike possessed a self-sufficient animal economy.

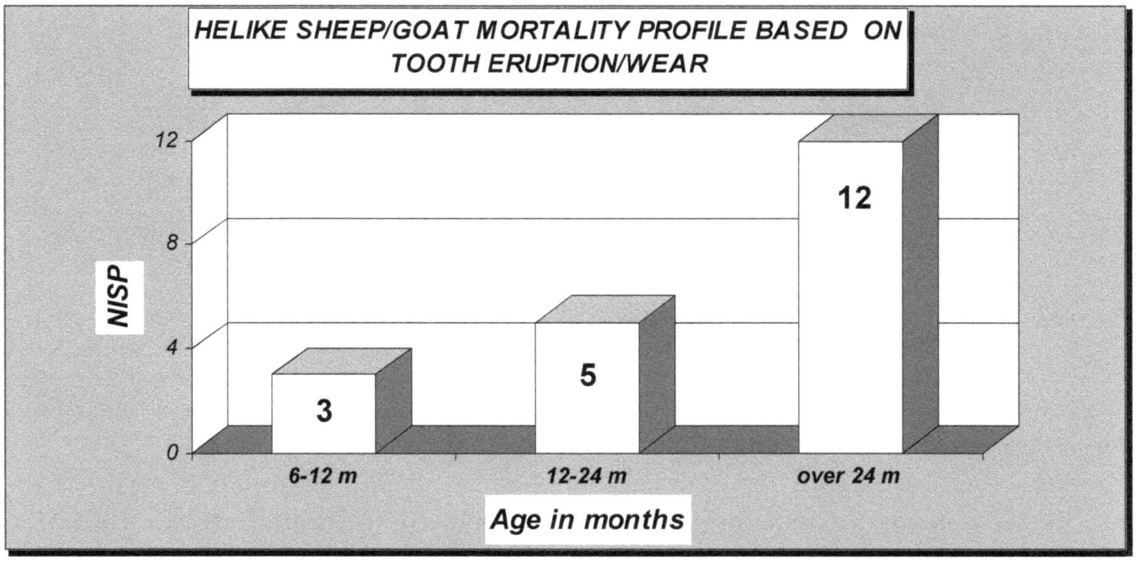

Figure 7.14 **(see appendix Table 10)**

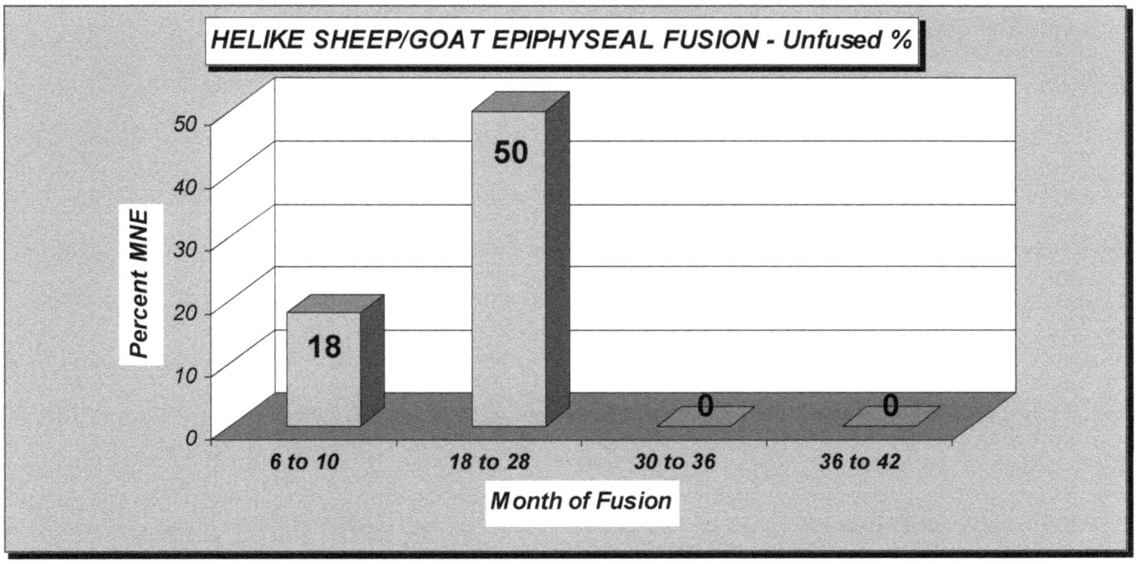

Figure 7.15 **(see appendix Figures 12-13)**

Both evidence of burning and cut marks appear on ovicaprid remains. Roughly seven percent of the total NISP for sheep/goat shows evidence of butchery. When viewed as a percent of MNE, the figure increases to 11 percent. These percentages are not significantly different from those obtained from pigs and cattle. As with the other domesticates, cut marks appear on various skeletal elements (*Figure 7.16*). Analysis of the distribution of the marks once again suggests primary butchery and disarticulation, as their placement clusters around areas of skeletal element articulation.

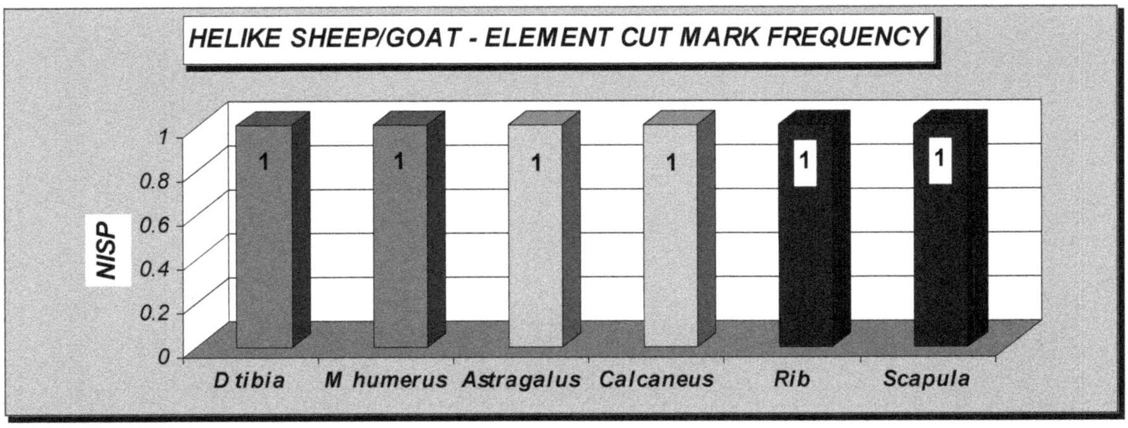

Figure 7.16 (see appendix Table 14)

Burning is present on 13 percent of the assemblage (NISP) or 20 percent (MNE). The majority of the burned bone appears to be concentrated in one particular area, which may have been a hearth and/or food preparation area. Further excavation is needed to clarify this hypothesis, however. As with evidence of butchery, burning occurs of a variety of elements. No fragment bears both cut marks and evidence of burning.

Summary:

The faunal assemblage at Helike suggests a largely self-sufficient settlement with a small, local animal economy. Analysis of the skeletal material from the major animal domesticates (pig, cow, sheep/goat) indicates that these species were most likely being raised and consumed locally. Skeletal element frequencies, along with mortality profiles, show individuals of most ages and body parts. The one exception to this generalization is with regard to pigs. Individuals of 12-24 months of age appear to be conspicuously absent at Helike (when aged on epiphyseal fusion), and this may suggest the possibility that Helike was rearing pigs and providing them to nearby settlements. This suggests inter site trade relationships, and is an issue that will be addressed in Chapter 8.

While the species composition at Helike is comparable to other Early Bronze Age sites in Greece, and those under analysis in this study, the relative percentages of the individual species are unique. Cattle and pig bones made up the majority of the faunal assemblage, followed closely by sheep/goat. In general, data from contemporaneous sites usually suggests that the heaviest exploited species are ovicaprids. Helike is unique in that pigs are the heaviest exploited species. Chapter 8 will address possible theories stemming from this unique pattern.

II. Tsoungiza

Physical and Geographical Location:

The Early Bronze Age site on Tsoungiza Hill is situated in the village of Ancient Nemea (Herakleion) in the northeast Peloponnese (*Figure 2.1*). It lies on the old road between Corinth and Argos, roughly 40 km south east of modern and Ancient Corinth. Tsoungiza Hill is inland, at the head of the Nemea Valley, within two hours walking distance to the nearby site of Mycenae. The geographical location provides fertile lands and riverine access, and although inland, it is located within a few hours from the water of the Saronic Gulf.

Tsoungiza was excavated as part of the Nemea Valley Project (NVAP) (Pullen 2006, in press). Excavation on the Bronze Age settlement occurred in the 1920s by J.P. Harland and in 1984-1986 under the direction of James Wright. The project was sponsored by Bryn Mawr College under the auspices of The American School of Classical Studies at Athens (ASCSA). The prehistoric settlement of Ancient Nemea on the hill of Tsoungiza contains remains spanning the Early Neolithic (roughly 6th millennium BCE) through the end of the Late Bronze Age (1200 BCE). The Early Bronze Age phase of Tsoungiza encompassed roughly the entire third millennium BCE, and it is this phase with which this study is concerned.

The Early Helladic Settlement:

The Early Helladic settlement at Tsoungiza is comprised of several buildings, some of which were two-storied, and it is from within these, and several pits and cisterns, that the animal bones were recovered. Tsoungiza possessed substantial architectural remains, especially during the Early Helladic II phase of the site (2400 BCE). Excavation of building A revealed walls over 1 m thick and has been reconstructed as having possessed a second story. Within this structure was found jewelry, a lead seal, and Cycladic pottery[46], suggesting that Tsoungiza was linked to the wider world through trade.

The Faunal Remains:

The faunal remains from Early Bronze Age Tsoungiza were analyzed by Paul Halstead, who graciously made his data available to me. They represent the results of excavations from 1984-1986 seasons. The material comes from a variety of contexts – and

[46] As Harland dug the site in the 1920s, including House A, there is little from this structure. However, plentiful pits and cisterns, as well as the large outdoor surface surrounding House A, were extant from which came the majority of faunal remains (Pullen, pers. comm.).

the spatial analysis of the bones was accounted for in quantification. The faunal remains from Early Helladic Tsoungiza have been analyzed together with those originating from the Final Neolithic. As a result, the collection under consideration will be larger than that originating from just the Early Helladic phase of the site. Since distinction was not made between the two periods during analysis, it cannot be made here. It must also be noted that site occupation at Tsoungiza encompassed more time periods than at Helike. At Tsoungiza, the faunal assemblage under consideration spans four periods (Final Neolithic, Early Helladic I, II and III). Further, within each of these chronological periods exist several occupation phases. Since only a portion of the data were broken down according to these chronological periods/and or phases of occupation, it was not possible to extract only those faunal remains from Early Helladic II/III contexts. Therefore, it must be noted that the assemblage from Tsoungiza will be larger than Helike, and moreover, may contain inherent difficulties for comparative study. As at Helike, the faunal remains from Tsoungiza yield evidence for heavy exploitation of three main species: cattle (*Bos Taurus*), pig (*Sus domesticus*), and sheep/goat (*Ovis aries/Capra hircus*). Pigs account for 38 percent of the identified species, cattle 14 percent and sheep/goat 43 percent. Roughly five percent of the total assemblage is comprised of other species (*Figure 7.17*). The remaining identified species consist of dog (*Canis familiaris*), auroch (*Bos primigenius*), boar (*Sus scrofa*), red deer (*Cervus elaphus*), roe deer (*Capreolus capreolus*), fox (*Vulpes vulpes*), hare (*Lepus europaeus*) and tortoise (*Testudo sp.*) (Halstead 2006, in press).

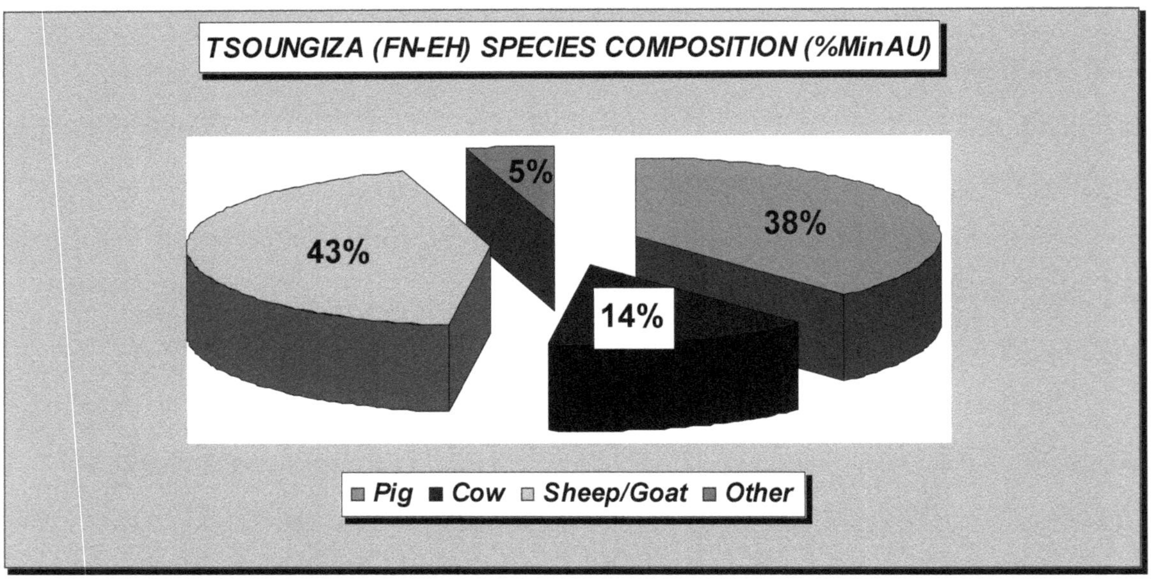

Figure 7.17 **(see appendix Table 5)**

Methodology:

The majority of the faunal remains analyzed from Early Bronze Age Tsoungiza were recovered in the trench by hand, and this was supplemented by dry-sieving all of the deposit through a 5-7 mm mesh. A portion of the material was also wet-sieved through a 0.5 mm mesh. Bone was bagged in the field and analyzed in the Nemea Museum. Only that material that was identified to both anatomical part and taxon was recorded in detail, although all fragments were retained. In addition to identification of anatomical part and taxon, where appropriate the following variables were recorded: side of body, presence/absence of proximal and distal units, state of epiphyseal fusion/dental development, sex, pathological traces, fragmentation, traces of gnawing and burning, butchery marks, metrical data, and method of retrieval (i.e. trench or wet-sieve). Distinction between sheep and goat were made where possible, however much of the material was grouped into one category, ovicaprid. Measurable cattle and pig bones were too few for systematic separation into wild and domestic populations, and so this distinction was rarely made. However, available metrical evidence that was present suggested that domestic cattle and pigs predominated heavily over their wild relatives (Halstead 2006, in press).

Age at death of domestic mammals was estimated at Tsoungiza on the state of dental eruption and wear of mandibular cheek teeth (following Bull and Payne 1982; Deniz and Payne 1982; Grant 1982; Grigson 1982; Halstead 1985; Payne 1973). Age at death was also estimated using post-cranial bones on the basis of epiphyseal fusion (following Silver 1969). Metrical data was obtained according to von den Driesch (1976). Where possible, pelves were sexed on morphological grounds following Grigson (1982) and Boessneck et al. (1964).

Quantification:

The faunal assemblage at Tsoungiza was quantified in terms of minimum numbers of anatomical units (MinAU), after a method described by Halstead (1985). MinAU is the basic method of quantification used at Tsoungiza in estimates of the relative abundance of different body parts. It is roughly equivalent to MNE (Halstead, personal communication). MinAU was estimated as follows:

> Where two or more fragments might be derived from the same anatomical unit (e.g. a single left proximal tibia) of the same individual animal, only the most complete example contributes to the minimum number of anatomical units. Similarly, to simplify comparison between species with different numbers of foot bones, quantification of fragments of metapodial bones and phalanges has been standardized in terms of minimum numbers of feet: thus if two specimens of phalanx 2 of, say, sheep (or sheep/goat) could be derived from the same foot, only one contributes to the

MinAU. Assessment of MinAU was based on visual comparison of specimens and involved the strewing of anatomical/taxonomic groups (e.g. pig humeri) into sub-groups (left/right, proximal/distal, medial/lateral, used/unfused etc.) (Halstead 2006, in press "Faunal Remains FN-EH Nemea Tsoungiza").

Size and overall composition of the Early Helladic assemblage:

The faunal assemblage from Early Helladic Tsoungiza was subject to a certain amount of gnawing by pigs/dogs, and this played a significant role in shaping the fragmentation of the overall assemblage. This said, the majority of the bones have been interpreted as representing material discard by humans after some form of carcass processing. A high incidence of old breaks suggests possible marrow extraction by humans. The remainder of the fauna from Tsoungiza will be presented together, broken down generally into a discussion of mortality profiles and body part representation/skeletal element frequencies. Pigs will be discussed in greater detail in the following chapter.

Skeletal Element frequency/body part representation:

In order to analyze body part representation at Tsoungiza, skeletal elements from the major taxa were grouped into four larger categories:

- cranial (horn core, mandible),
- axial (pelvis, rib, sacrum),
- appendicular (scapula, humerus, radius, ulna, femur, tibia) and
- foot (metacarpal, metatarsal, phalanges).

Grouping the skeletal elements into these broader categories facilitates comparison between the taxa, and it also provides a graphic representation of the body parts recovered from each species. Figure 7.18 illustrates that all of the major skeletal elements are present for each of the domesticates, suggesting, as at Helike, local production and consumption of cow, pig and sheep/goat.

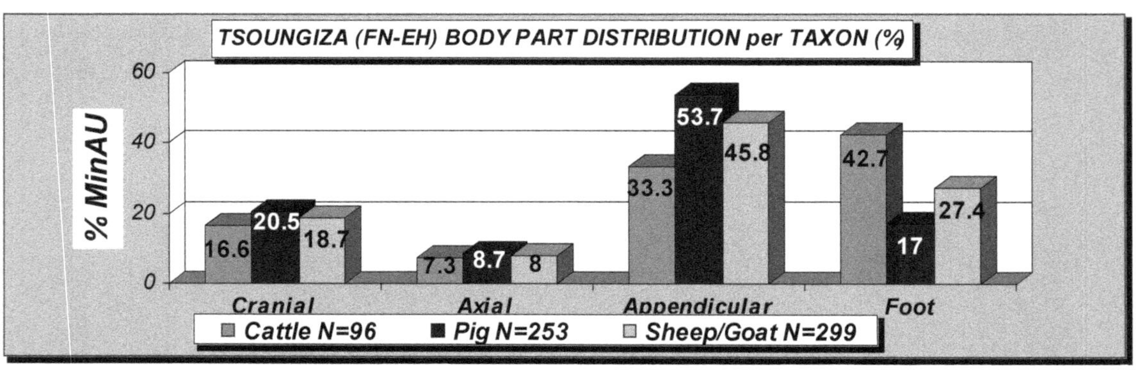

Figure 7.18 **(see appendixTable 7)**

A clearer understanding of Tsoungiza's animal economy necessitates an examination of the data provided by skeletal element frequencies against mortality profiles. Mortality profiles were generated using data from both month of post-cranial epiphyseal fusion, and tooth wear/eruption stages. Data generated from both methods corroborate one another's findings (*Figures 7.19* and *7.20*). With respect to pigs, the data indicates that the majority of ageable individuals were killed before their 24th month. Pigs have a young mortality profile at Tsoungiza, and given that pigs are not generally exploited for secondary products, these data suggest primary use as meat-producing animals. This pattern is also predicted by the high reproductive rate of this species. Dental data supports this age structure.

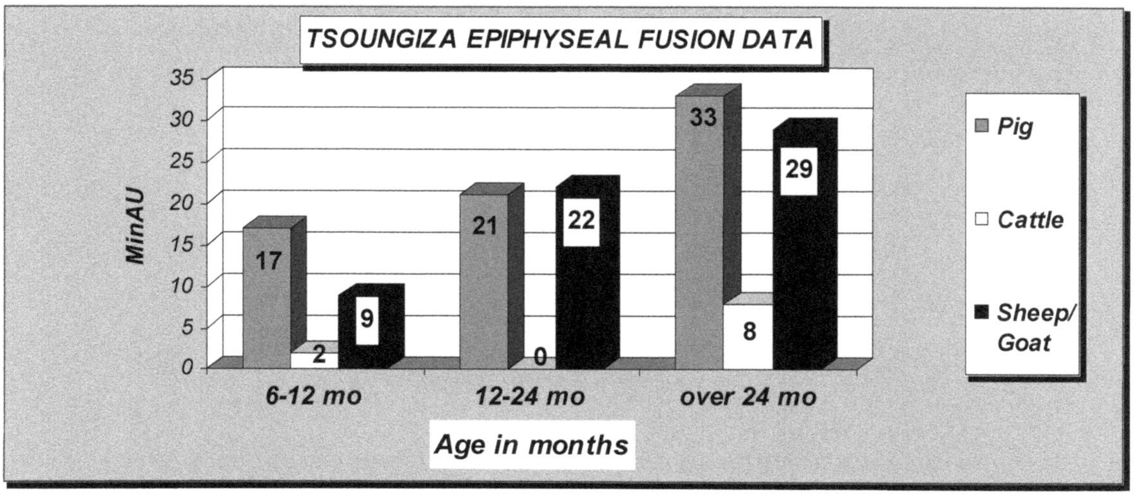

Figure 7.19 **(after Halstead 2006, in press)**

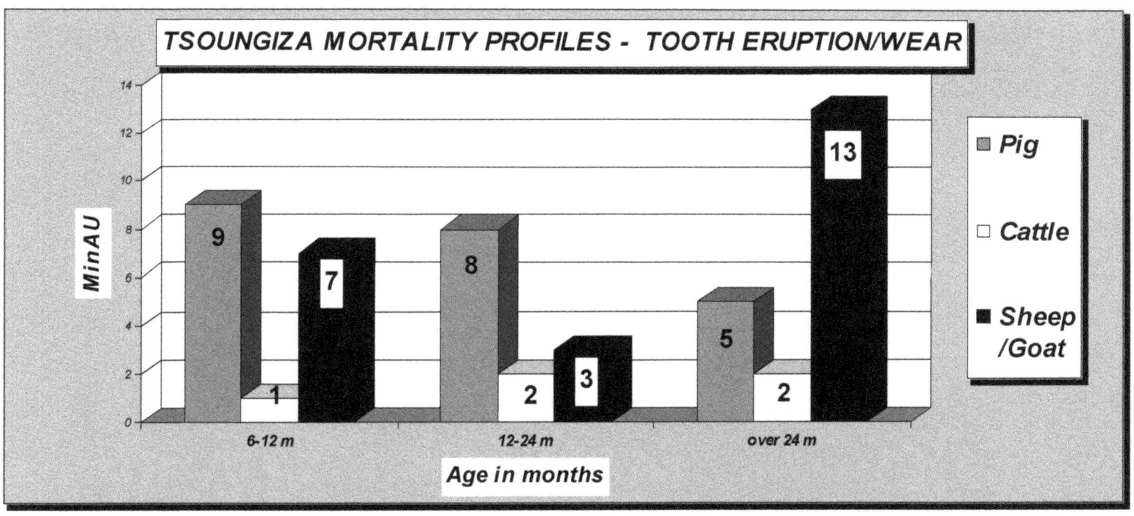

Figure 7.20 **(after Halstead 2006, in press) (see appendix Table 12)**

While little information can be gleaned from the data offered by cattle remains due to the very small ageable sample size, data from sheep/goat offer sufficient data to reconstruct their probable function in Tsoungiza's animal economy. While individuals under 24 months of age are present in the faunal assemblage, an almost equal number of individuals over two years of age are also present. This data is consistent with the utilization of sheep and goat for both primary and secondary products. While younger individuals could have served a primary function as meat producers, older individuals were most likely kept for the production of wool and dairy products. There thus appears to be a balance between the utilization of ovicaprids for both primary and secondary products. Again both dental data and data derived from month of epiphyseal fusion are in agreement for this species.

Halstead's data also offers information and analysis of probable butchery patterns for the major domesticates at Tsoungiza. He concludes that butchery marks and patterns of fragmentation provide evidence from the entire working sequence of carcass processing, from skinning to marrow extraction (Halstead 2006, in press). Thus, Tsoungiza also appears to have a rather self-contained animal economy that included production and consumption of the major animal domesticates.

Summary:

The faunal assemblage from FN-EH Tsoungiza is one of the first from Early Bronze Age Greece to have undergone rigorous, standardized methods of retrieval and analysis. In general, the assemblage was mediated by carnivore activity and trampling, and unlike Helike, may have been exposed to the elements for a longer period of time, thus affecting the final taxonomic and element composition. While Halstead argues that the Tsoungiza assemblage is too small for detailed consideration of animal management practices, it offers valuable comparative information that will form the start of an ongoing conversation. This is an issue that will be addressed in-depth in the following chapter.

While the methods utilized at Tsoungiza will hopefully serve as a model for the future of Greek faunal studies, both Lerna and Tiryns were not subject to such rigorous and thorough analysis. This renders inter site comparisons difficult, and it is to these sites that this chapter now turns.

III. Lerna

Physical and Geographical Location:

Lerna is situated in the Argolid, in the eastern Peloponnese, in the modern village of Myloi. It lies on the western shore of the Bay of Argos, directly across from the contemporaneous sites of Asine and Tiryns. To its north is the Saronic Gulf, a body of water that would have supplied easy access to eastern trade routes. As with Helike, there has been significant change in the location of the modern coastline, and while Lerna is currently nearly one km inland, it was probably a port city during the Bronze Age settlement. Lerna lies in close proximity to Tsoungiza, and is roughly 60 km from modern and Ancient Corinth (*Figure 2.1*).

J.L. Caskey described the settlement at Lerna as: "A low rounded hillock stand[ing] by the south bank of the stream Amymone, which runs its short course from the Lernaean spring to the Gulf of Argos (Caskey 1954:3). Lerna is currently situated in fertile plains, amongst modern vineyards, with a view to the east of the sea and to the west of low hills gradually turning into the rocky mountains that comprise much of inland Greece. Lerna is at the mouth of a route over the mountains into Arcadia. Sea contact is easy with the Peloponnese and with the Argolid Peninsula, which separates the Gulf of Argos from the Saronic Gulf. Contact with the islands of Keos, Melos, Thera, and the other Cycladic Islands, and by this means either with Crete to the south or with Anatolia to the northeast, would also have been possible. The site spans a time frame from the Neolithic to the Roman period. Caskey's excavations concentrated on Bronze Age remains, primarily Middle Helladic II and III (ca. 2000-1600 B.C.). The Bronze Age settlement encompassed an area roughly 180 meters from east to west, and 160 meters from north to south. This study will focus on the Early Helladic faunal remains only (Lerna III/IV).

The Early Helladic Settlement:

Excavation at Lerna spanned the period from 1952-1958. The excavations were under the direction of J.L. Caskey of the University of Cincinnati, under the auspices of the American School of Classical Studies at Athens (ASCSA). According to Caskey (1969:iii), Lerna appears to have been abandoned at the end of the Neolithic and remained unoccupied throughout the first stage of the Early Bronze Age, Early Helladic I. The settlement appeared to reach its height during the Early Helladic II phase, which at Lerna is divided into four main phases.

Early Helladic II at Lerna (Lerna III) was witness to significant rebuilding in comparison with earlier periods, and at this time the settlement was fortified with a circuit wall and towers (Caskey 1969:iii). Larger buildings, including the well-known House of the Tiles, date to this period. The House of the Tiles is rectangular in shape, with two large and several smaller rooms, and walls almost a meter thick (Caskey

1955:40). Staircases could be ascertained on each of the long sides of the house indicating the presence of a second story (Caskey 1955:39). Radiocarbon dating indicates that the House of the Tiles was destroyed by fire ca. 2200 B.C. (Caskey 1969:iii).

While Caskey referred to the House of the Tiles a "palace", this term is most likely a misnomer. During that time a palace would probably function as a multi-purpose building, more akin to an administrative center. It may have functioned as a decision-making center, warehouse and distribution center, or even living quarters for those of higher social status. Adding to the allure of the House of Tiles are more than 150 clay sealings, found in one of the side rooms. These sealings have been the subject of much study (see discussion in Chapter 2), but in brief have been interpreted as evidence of trade. They lend support to the hypothesis that the House of the Tiles may have been an administrative center, and Lerna may have played a wider role in Early Bronze Age trade. Lerna may have been exporting its own products and the seals indications of the producers or manufacturers. Alternatively, the seals may have belonged to containers originating from elsewhere, in which case Lerna may have been acting as a central distribution point or trading center. These ideas have been discussed at length in chapter two.

The Early Helladic III period (Lerna IV) is also witness to building and site occupation, although on a smaller scale than the previous period. The House of Tiles was never rebuilt, and there is no evidence for large scale architecture. It appears that Lerna may have experienced depopulation, and this may have led to her eventual abandonment. There is diminished evidence for trade at Lerna IV – a phenomenon that is common during the Early Helladic III period in the Peloponnese. Carbon-14 dating has placed the Early Helladic III period to the close of the third millennium and beginning of the second millennium.

The Faunal Remains:

The faunal assemblage at Lerna was published by N.G. Gejvall in 1969, and is the result of a preliminary field study of the material undertaken in 1958, at the project storerooms in the museum of Ancient Corinth. The faunal remains from Early Helladic II-III Lerna (Lerna III/IV, unmixed layers) are comprised of 4738 fragments (NISP). During the summer of 2005, I obtained permission to re-analyze some of the fauna from Lerna held at the archaeological museum in Argos, Greece. This material is currently under study by David Reese. I had intended to compare my data with Gejvall's (1969) published material, however this approach ended up being problematic. I found that the material held in Argos was not part of the published assemblage. Although Gejvall had examined the material, it was not possible to ascertain why these bones were not included in the publication. The time spent in Argos did prove useful, however, in gaining an idea as to the nature of the assemblage with regard to taphonomy and general condition of the specimens. All pig data stored in Argos was recorded and cut mark casts and photos taken. Future study is intended on the cut marks, although they are not analyzed in this study. In sum, the figures given for the faunal data from Lerna will be those of Gejvall's (1969) published analysis from the unmixed layers at Lerna III and IV (EH II/III).[47]

The faunal assemblage at Lerna more closely resembles that from Tsoungiza than Helike in regard to the relative proportions of species present. At Lerna, sheep/goats are the species represented in the greatest number (35 percent), followed by pigs (30 percent) and cattle (28 percent) (*Figure 7.21*). Gejvall (1969) concluded that the faunal assemblages at Lerna suggested animal husbandry with full domestication of the major species. He concluded that by Early Helladic II breeding of domestic pig, sheep/goat, and cattle was occurring. The donkey was also introduced at this time.

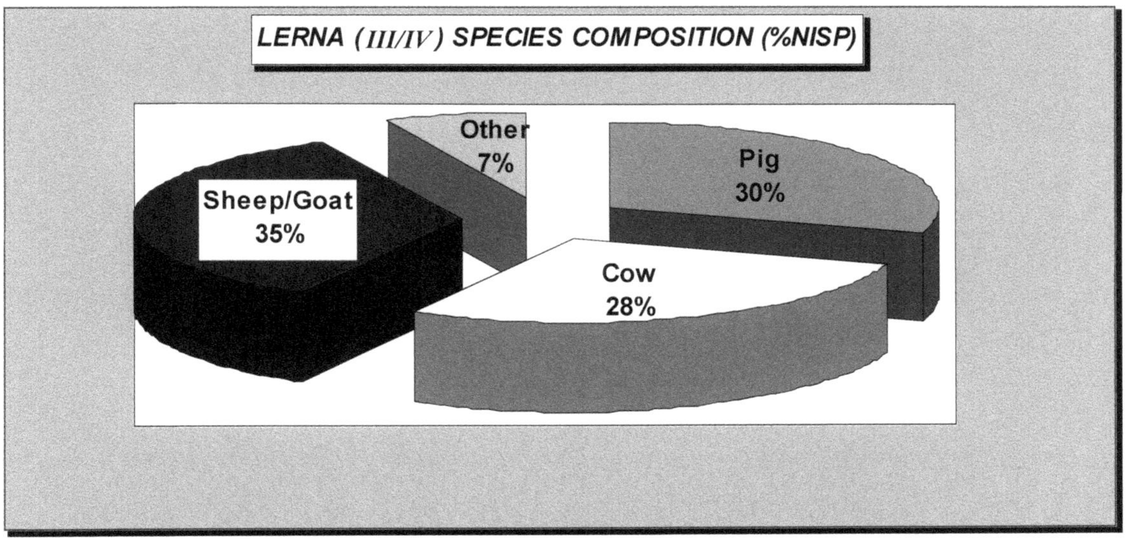

Figure 7.21 **(see appendix Table 5)**

In addition to the major animal domesticates, badger, common otter, and beech martin appear in Lerna IV (Gejvall 1969:40). A number of avian fauna are also present, including heron; mallard; gray lag goose; whooper swan; crane; rock partridge; pigeon; eagle owl, the largest of the European owls; raven; and hooded crow (Gejvall 1969:48).

> ...bird bones from the periods earlier than Lerna IV include only swimming birds and waders plus one raptor (goshawk), whereas the later periods include a series

[47] The raw numbers used can be found in Gejvall (1969:10), Table 6. They do not include *Aves or Pisces.*

of additional species of typical rock (and steppe [sic]) dwellers and some domestic birds (?[sic]) as well (Gejvall 1969:49).

Gejvall (1969:48) concluded that the type of fowl extant during this period suggest an environment possessing shallow water (duck), muddy shores and pasture land close to water. This is consistent with hypothetical reconstructions of the area.

Methodology:

The faunal assemblage from Lerna was analyzed in the storerooms of the Museum in Ancient Corinth. Gejvall was not in the field at the time of excavation, and relied on excavation notes to augment his analyses. Preliminary analysis consisted of visual inspection of every bone fragment collected in the field. The exact collection methods at Lerna were not discussed in Gejvall's (1969) publication, and thus the assumption is made that collection was carried out primarily by hand in the trenches. The assumption is also made that sieving did not occur. The following is a brief synopsis of Gejvall's (1969) analytical procedure.

All bone fragments were initially sorted into two main units: those that were too small for measurement (splinters) and those that appeared measurable. Bone was bagged in the field by excavation lot number, and these bags were subsequently assigned a unit number. A card was made for each unit number, and on this card preliminary notes were recorded using a mnemotechnical code (system of abbreviation) detailed in the published volume.[48]

These cards contained a variety of information from which Gejvall's methods can be reconstructed. The main categories of data recorded were: species of animal, number of fragments of the species, element, and side if ascertainable. Other "characteristics" recorded were erupted teeth and stage of eruption, abrasion and abnormal number of teeth, cut/gnaw marks, burning, coloring by metal, pathology, shed antlers, etc. Ageing data was compiled solely on dentition, specifically on tooth wear and state of eruption of mandibular teeth. Determination of sex was limited to that gleaned from horn cores, and thus is not available for each species. All measurable fragments and other fragments and samples which were to be saved for later investigation were numbered in black ink (Gejvall 1969:2). Fauna from Lerna were then processed on computer with an early form of coding lists and punch cards, in a system that also took into account find contexts and spatial distribution of the bones.

Although admittedly ahead of its time in the scale of analyses, the methodology employed at Lerna does pose some problems. It is unclear how "measurable" and

[48] These cards still exist, and David Reese has been examining them in his re-analysis of the Lerna fauna.

"unmeasurable" fragments were determined. The numbering system is also unclear, as the fauna I examined held at Argos did not bear any numbers.

Quantification:

Gejvall presented the fauna in terms of "units" (NISP), the total number of which amounted to 15,621 for all chronological periods and trenches. In addition to providing raw data units, the terms BIND (coherent sample of fragments belonging to one individual) and MIND (minimum number of individuals) were used. Gejvall's MIND is equivalent to MNI. He does not provide detailed information on how BIND or MIND were calculated, and thus it is not possible to ascertain from his MIND counts whether they were minimized or maximized. When data from Lerna is used in this analysis, note is made whether counts are in terms of NISP or MIND (MNI), with NISP used whenever possible to facilitate inter-site comparisons.

Size and overall composition of the EH assemblage:

Gejvall structured his faunal analyses by species, breaking down his discussion according to each of their roles diachronically at Lerna. This summary will provide an overview of his data as it pertains to just the Early Helladic II and III periods (Lerna III and IV).[49] Specifically, the numbers used for comparison here are taken from the unmixed layers III and IV at Lerna – or more simply, those layers that could be definitively attributed to the Early Helladic II and III periods. A total of 4738 identified specimens (NISP) were recovered from these two chronological periods. In addition to the major domesticates (pig, cattle, sheep/goat) (*Figures 7.22, 7.23*), dog, fox, wolf, rabbit, red deer, roe deer, donkey, pine martin and otter were found. As previously mentioned, several avian species were identified, as were fish and mollusks.

[49] In view of the problems that may arise due to different methods of quantification, and different chronological periods at each site, the ensuing analyses will attempt to standardize the assemblages under consideration through the use of percentages rather than absolute numbers. It is acknowledged that the final analyses may be subject to a certain amount of oversimplification resulting from the comparison of various chronological periods and habitation phases; however this is unavoidable due to the nature of the published material. Every effort will be made to take this potential issue into account and analyses will attempt to mitigate potential problems.

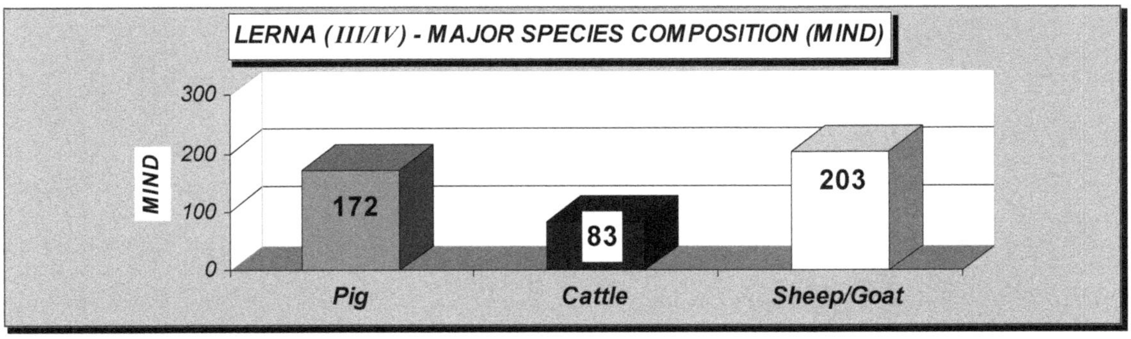

Figure 7.22 (Gejvall 1969:19-30)

Faunal analyses at Lerna subdivided the major species into wild and domestic forms. This subdivision was often based on physical observation, and due to the position adopted by this research that the wild/domestic dichotomy is often riddled with problems, this subdivision was abandoned. Thus, where Gejvall has created two categories for pig, *Sus scrofa* and *Sus transitional/domesticus*, this study groups these two divisions into one, *Sus scrofa* (pig). Further distinction was made between sheep and goat, where possible, as well as *Bos primigenius* and *Bos transitional/domesticus*. Again, these categories have here been combined.

Taphonomic factors may have also played a large role in the formation of the faunal assemblage at Lerna. A significant number of dog remains are present, suggesting the strong possibility that the assemblage as whole may have been subjected to carnivore gnawing. The degree of fragmentation of the assemblage can be hypothesized by examining the NISP/MNI ratio (units/MIND) (Lyman 1994:332). When calculated, this ratio suggests a highly fragmented assemblage. With respect to pigs, the ratio is 1432:172 or 8.3, the ratio for cattle is 1322:83 or 15.9, and sheep/goat is 1637:203 or 8.1. These ratios indicate that cattle bones were subjected to the greatest amount of fragmentation, followed by pigs and then sheep/goat.

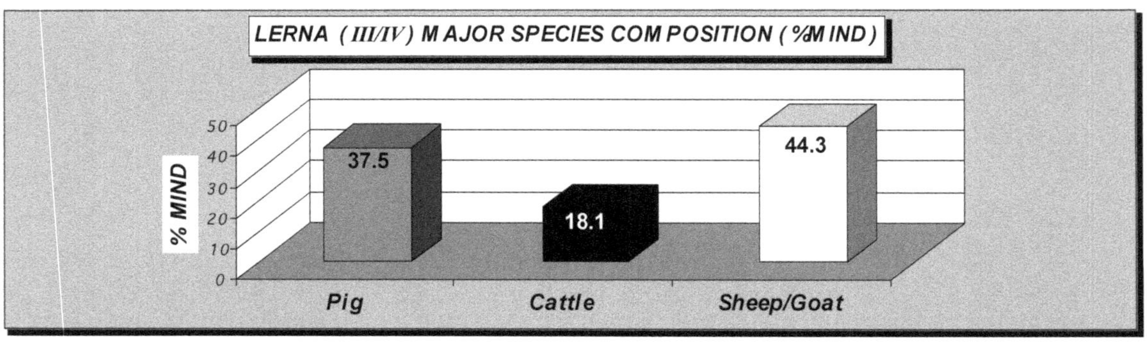

Figure 7.23

Skeletal element frequency/body part representation:

Analysis of the major body parts of each of the major species at Early Helladic Lerna suggests that, as at Helike and Tsoungiza, the animals were being raised and consumed on site. Skeletal elements subdivided into four main categories to facilitate comparison with other sites, and provide a general picture of animal exploitation on site. These categories are comprised of the following elements:

- cranial (horn core, mandible, maxilla, cranium),
- axial (vertebra, ribs, sacrum, pelvis),
- appendicular (scapula, radius, ulna, humerus, femur, tibia, fibula, patella), and
- foot (metacarpal, metatarsal, phalanx, astragalus, calcaneus).

Cursory examination of Figure7.24 illustrates a high frequency of cranial and appendicular elements, suggesting on site mortality.

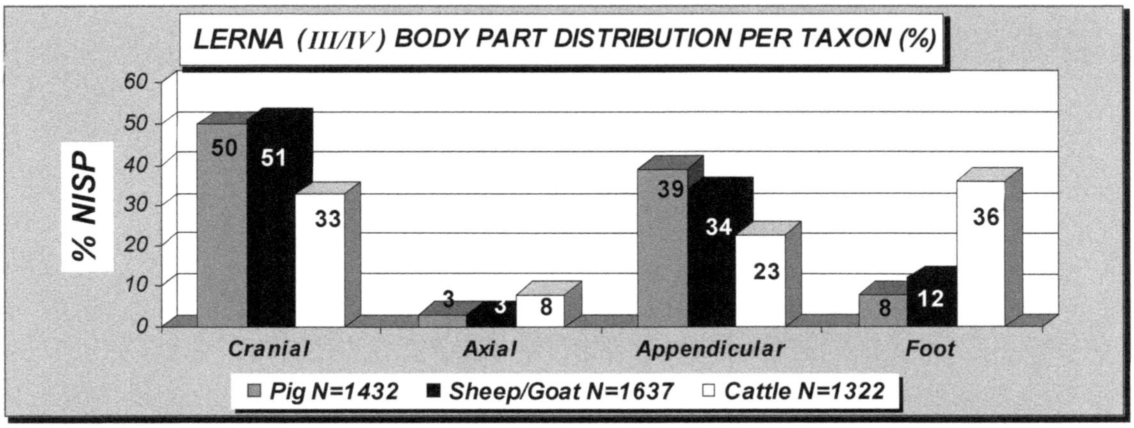

Figure 7.24 **(see appendix Table 8)**

Mortality profiles:

Gejvall (1969) analyzed the ages at death of the species at Lerna based on dental eruption/wear. The published report provides no discussion of mortality based on epiphyseal fusion. Data from both Helike and Tsoungiza suggest that the two methods mirror one another closely enough that dental data used on its own still provides strong information on the age at death of the animals in the assemblage. Examination of this data from Lerna shows marked differences from Helike and Tsoungiza. For each of the major species, mortality appears to have been after the 24th month of life (*Figure 7.25*). With respect to sheep and goat, this data is consistent with an economy relying heavily on secondary products. Regarding cattle, the data suggest use as traction animals and also perhaps secondary products (though in the absence of sexing data this hypothesis becomes difficult to support).

Mortality profiles at Lerna are most surprising for pigs. Over 60 percent of the pigs appear to have lived past the two year mark. While younger individuals are also present, this is an unusually high percentage. Analysis and discussion of this trend will occur in the following chapter.

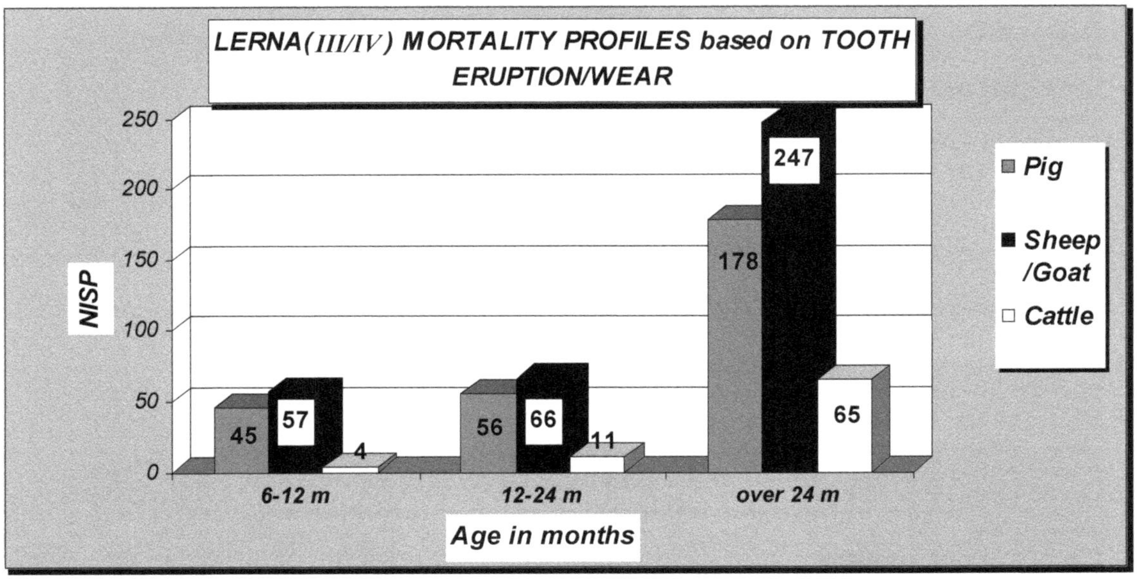

Figure 7.25 **(after Gejvall 1969) (see appendix Table 11)**

Summary

In Gejvall's (1969:51) summary of the fauna from Lerna, he concluded that cattle played a major role as meat producers at Lerna. Although their numbers are fewer than pigs and sheep/goat, he placed importance on the volume of meat produced by this animal, in comparison with the other domesticates. Due to the multi-phase nature of the site, a large portion of his analysis rests on examining changes in species composition through time. Attention is given to the changing ratios of wild vs. domestic animals. Gejvall views the shift toward a higher percentage of domesticated species, in conjunction with mortality profiles that indicate an older age at slaughter, as evidence for the development of a breeding and keeping program of animals at Lerna. He also allowed for the possibility that these changes may have been the result of a changing biotope (Gejvall 1969:53).

The faunal assemblage from Lerna is important as one of the first zooarchaeological analyses in Bronze Age Greece. The data provide an important body of evidence on which to draw comparisons with the new data from Tsoungiza and Helike. The following discussion of Tiryns will contribute further to this corpus.

IV. Tiryns

Tiryns is best known as one of the great Mycenaean period sites in the Argolid, forming a triumvirate of sorts with Mycenae and Midea. Located a few kilometers inland from the Argolic gulf, just on the edge of the modern city of Nauflio in the eastern Peloponnese (*Figure 2.1*), the acropolis of Tiryns (as it is sometimes called) occupies an impressive position on high ground. Its location affords it a strategic view of the surrounding countryside, and the waters of the Saronic Gulf. The site that greats visitors today is remarkable for it's amazingly thick and massive Cyclopeaen fortification walls (4.5 – 7 m thick) , however these architectural remains are the result of the Late Bronze Age occupation.

Settlement at Tiryns stretches back to the Neolithic period (about 5000 B.C.). It was followed by successive settlements whose remains have been destroyed almost completely by the extensive construction of the later Mycenaean age. Enough evidence survived from the settlement of the Early Bronze Age (2500-2000 B.C.) to prove the existence then of a series of apsidal houses arranged around a very huge circular building (28 m in diameter) on the summit of the hill, often referred to as the Rundbau, which, following more recent excavation, has been interpreted to have functioned as a granary (see Chapter 2 for discussion). A significant problem confronting the understanding the Early Bronze Age period at Tiryns is that its remains lie under the later Mycenaean palace, and as a result, only a small portion of it has been excavated. Based on these excavations, however, it has been concluded that Early Bronze Age Tiryns may have been similar to Lerna IV (Hägg and Konsola 1986).

As at Lerna, Tiryns eventually became a fortified settlement, and although the extant walls date to the 13th century B.C. (Late Helladic IIIB), fortification may have existed earlier. After the disintegration of the palatial system (about 1200 B.C.), the Acropolis continued to be used mostly as a cult place. The site had become deserted when Pausanias visited it during the second century A.D. Archaeological excavation of ancient Tiryns is linked with the name of Heinrich Schliemann who in 1876 opened the first trenches in the Acropolis and the site outside the walls. Schliemann continued excavation in 1884/5, and from 1905 to 1920, excavation was conducted by the German Archaeological Institute. At the end of the 1950's, excavations in the area were undertaken under the supervision of the Greek Ephorate of Antiquities.

The Faunal Remains:

Study of all faunal remains at Tiryns was undertaken by von den Driesch and Boessneck (1990), and those belonging to the Early Helladic period (EH II-III) comprise a small portion of a larger volume on the Mycenaean settlement of Tiryns. The fauna were studied over several seasons during the late 1970's and early 1980's, and while detailed description of the collection of animal bone is not available, it is likely that collection methods at Tiryns were similar to those throughout Greece at the time of its excavation, namely hand collection in the field of visible specimens. It is assumed that sieving of any type was not carried out. Von den Driesch and Boessneck (1990) utilized a number of analytical methods in their analysis of the fauna, including (when possible) sex determination, ageing (based on mandibular teeth) and skeletal element frequencies. Their report contains detailed analyses on each of the major animal domesticates, coupled with extensive metrical data. In addition to cattle, sheep, goat and pig, dog, horse/donkey/mule and deer were identified. Distinction is made between wild and domestic fauna on morphological grounds.

Size and overall composition of the Early Helladic Assemblage:

The faunal assemblage from Early Bronze Age Tiryns is quite substantial, totaling 4443 identified specimens (NISP), Figure 7.26.

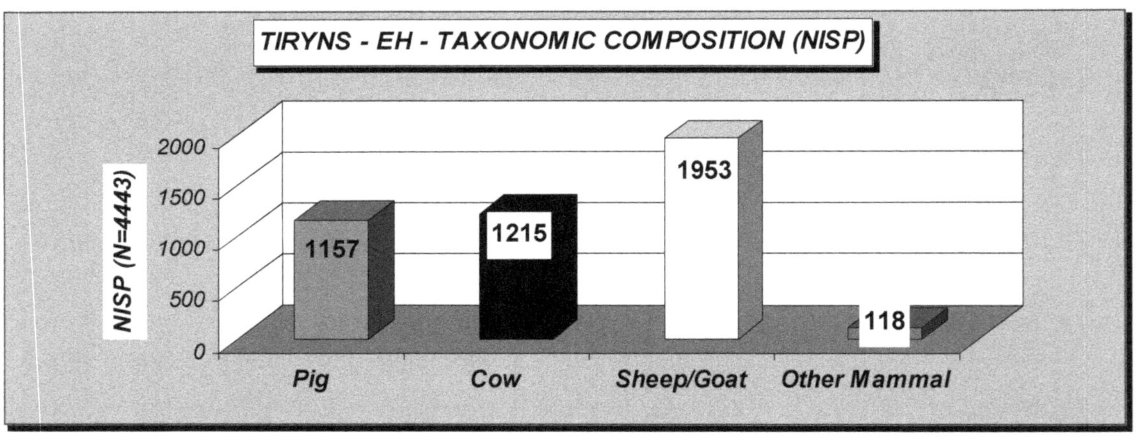

Figure 7.26 **(see appendix Table 5)**

Broken down into the major taxa, yields species ratios similar to those found at Tsoungiza and Lerna (*Figure 7.30*). At both Tsoungiza and Tiryns, sheep and goat account for the majority of the identified specimens (44 percent). However, Tiryns parallels Lerna in their similar ratios of cattle and pigs. At Tiryns, sheep and goat are followed by cattle (27.3

percent) and pigs (26 percent). Other mammals comprise 2.7 percent of the assemblage (*Figure 7.27*). Tiryns differs from Helike, Lerna and Tsoungiza with respect to the exploitation of pigs in particular, who here form the least exploited species. The authors, von den Driesch and Boessneck, suggest that the assemblage as a whole may have been subjected to carnivore activity, an issue which will be addressed shortly (1990:95).

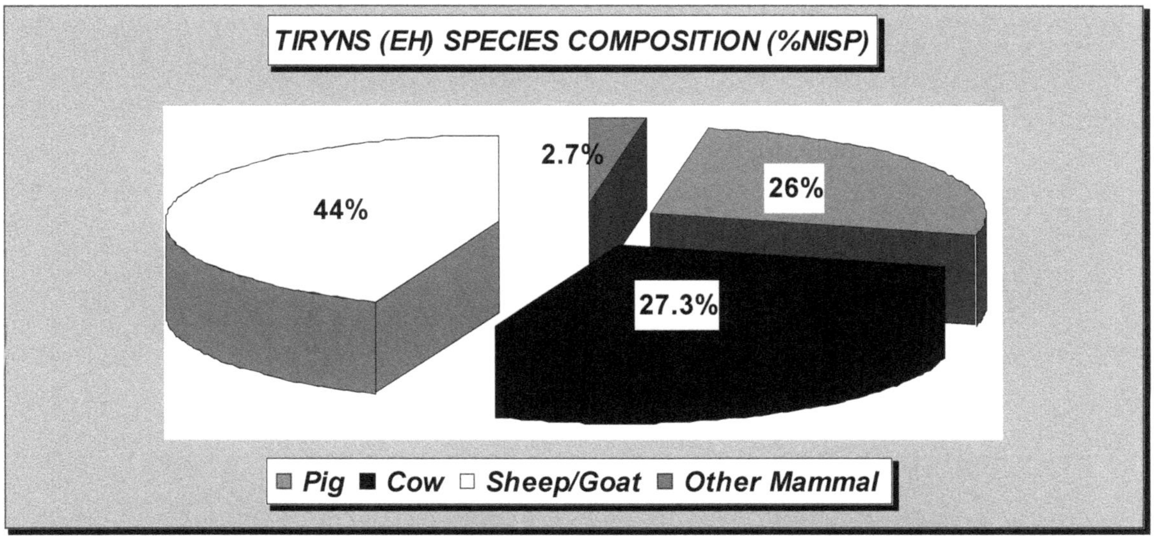

Figure 7.27 **(N=4443) (see appendix Table 5)**

Skeletal element frequency/body part representation:

Analysis of the distribution of skeletal elements for each of the major domesticates suggests that the animals were very likely raised in the vicinity of the settlement – if not in it. For each of the species, all body parts are present (*Figure 7.28*), however their relative frequencies differ markedly from each of the other assemblages under analysis in this dissertation (*Figure 7.31, page 163*). The main meat-bearing bones (limbs) form the largest group of skeletal elements recovered from Early Helladic Tiryns, followed by axial elements, the head and finally feet. These relative frequencies remain similar for each of the major species under study – and the paucity of foot bones is especially marked in pig.

Von den Driesch and Boessneck suggested that the pattern of skeletal element distribution was tied to two variables: transport and taphonomy (including carnivore activity) (1990:95). As discussion of the variables behind the distribution of skeletal elements pertains mainly to the Mycenaean phase of the settlement, it is difficult to ascertain whether explanations provided for the Late Helladic phase necessarily apply for the earlier settlement. Von den Driesch and Boessneck (1990:95) suggest that since the settlement lies on a hill, there would have been little room for animal husbandry, hypothesizing that most of the animal economy at Tiryns would have centered around the settlement in the immediate vicinity (but outside the city walls). They allow for the possibility that smaller animals, such as pigs and dogs, may have been kept within the settlement and thus slaughtered within the walls. They suggest that larger animals, such as cattle, would have been kept outside the settlement confines where they would have been slaughtered, and then transported into the settlement. They further suggest that if animals were brought into the settlement as a commodity, relatively worthless heads and feet would not be present.

However, cranial and foot bones are present for all of the domesticates – although in smaller quantities than the appendicular and axial elements. If transport and trade are not factors wholly responsible for this phenomenon it is possible that taphonomic factors could also have been at work. The presence of dog in the assemblage suggests that carnivore activity could have acted on some of the smaller, less dense bones, such as the foot bones of pigs – who were slaughtered young and whose bones would therefore be less dense (von den Driesch and Boessneck 1990:95). Dogs are known to completely destroy the smaller bones. It is also possible that the larger bones, which are fragmentary, were also subject to carnivore activity. Since quantification was done solely on NISP, there is no MNE or MNI data available. This makes determining the amount of fragmentation of the assemblage impossible. The authors do remark that the larger bones are often broken into smaller fragments, and suggest that this may have falsely shaped the skeletal element distribution – skewing it on the side of appendicular elements. While possible, examination of NISP/MNE ratios at Helike suggests that the relative skeletal element ratios are not significantly different when one method is used over the other, however this may simply suggest that Helike's assemblage is less fragmentary.

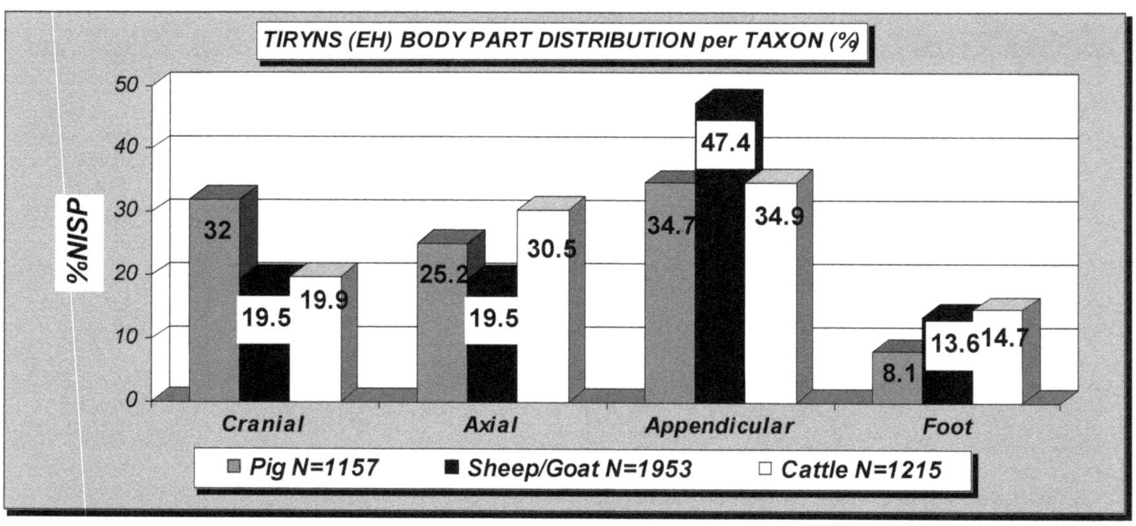

Figure 7.28 **(see appendix Table 9)**

Mortality profiles:

When possible, ageing of all the major domesticates was determined on the basis of mandibular tooth eruption – specifically molars and premolars. While the sample sizes for each species is relatively small, contrasting the mortality profiles for each may illustrate the different management strategies that were likely employed, and transitively, the animals' probable economic utility. Pigs appear to have been slaughtered during their prime – in the one to two year age range, however older individuals were also present, suggesting some local breeding. With the notable exception of sheep and goat, individuals under six months of age are absent from the assemblage (*Figure 7.29*). While taphonomic factors may be partially accountable for this distribution, the presence of some individuals in this age range suggests that economic factors, such as desire for meat, may have been responsible.

The limited data on species mortality suggests that pigs and ovicaprids may have served a primary role as meat producers. While not surprising with regard to pigs, the paucity of ovicaprids over two years of age is interesting. Von den Driesch and Boessneck (1990) suggest that ovicaprids had high adult survivorship, and perhaps this is simply not illustrated by dental ageing methods. If this suggestion was based on epiphyseal fusion data, they do not include it in their report. The older age of cattle may be a product of small sample size, and it may also indicate that heavy reliance was not placed on them as meat producers.

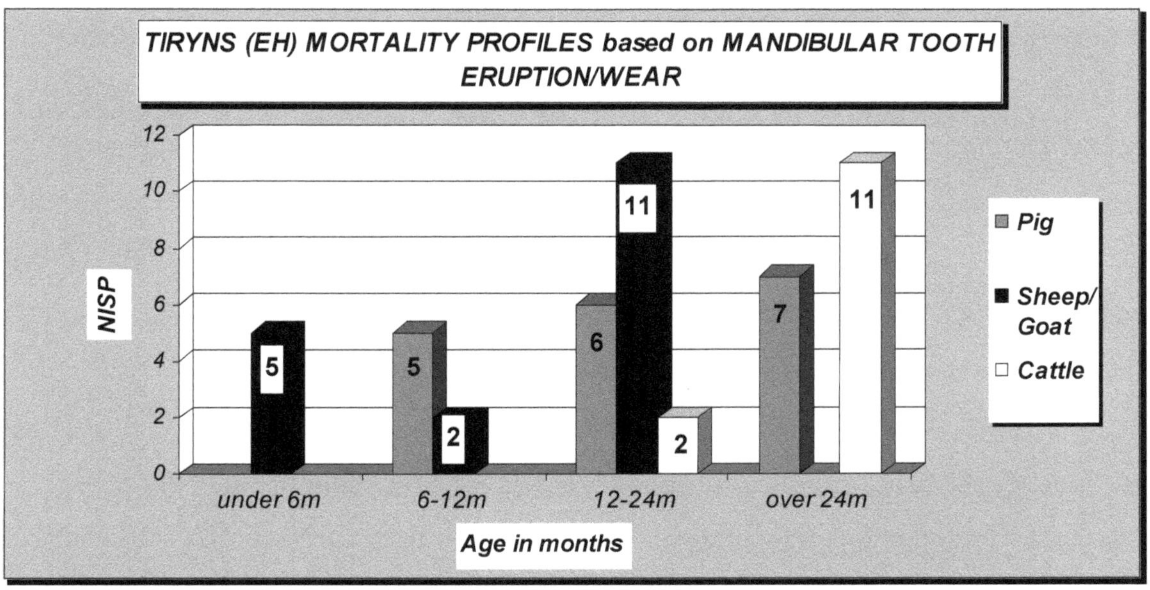

Figure 7.29 **(after von den Driesch and Boessneck 1990) (see appendix Table 13)**

Summary:

At Early Bronze Age Tiryns, hypotheses regarding animal utilization and the ancient animal economy can be drawn on species ratios and comparison with other Early Helladic faunal assemblages. Examining the relative percentages of the major species suggests that sheep and goat played a significant role Tiryns. According to von den Driesch and Boessneck (1990) adult survivorship was high in sheep and goats, supporting conclusions concerning their role in the production of secondary products. Although detailed age profiles are not available, comparison with Tsoungiza suggests that ovicaprids account for a similar proportion of the species exploited (44 percent and 43.6 percent respectively). The similar species ratio, coupled with higher adult survivorship, is consistent with a secondary products economy at Early Helladic Tiryns.

Data derived from the remaining major domesticates can also be examined and interpreted this way. Examination of cattle remains (27 percent) suggests a significant dependence on them, either as meat-producers, or as traction animals. If Tiryns was a larger, administrative type of settlement, with intensive agriculture in the fields surrounding the settlement (as suggested by the granary role of the Rundbau), cattle would be a desired commodity. The suggestion that Tiryns practiced more intensive agriculture is given additional support by examining the proportion of pigs in the overall assemblage. Pigs account for just 26 percent of the exploited species, the lowest frequency found at any of the Early Helladic sites under analysis. This suggests several possibilities – one of which is intensive agriculture. Pigs are destructive to crops, and unless contained, can do great damage to them. If a settlement has inter-site economic trade relationships, it is quite possible that in exchange for secondary products from sheep and goat, pigs could be obtained. While data suggests that there was an emphasis on young pigs at Tiryns (von den Driesch and Boessneck 1990), the presence of some older individuals does not rule out that the possibility that some breeding may have been on site – perhaps at the household level. This is an issue this study will address in depth in Chapter 8.

Although the Early Helladic phase at Tiryns has not been extensively excavated, the evidence suggests that Early Bronze Age Tiryns was a substantial settlement of some importance. The large quantity of animal bone recovered from such a small excavated area, the species compositions, and the existence of large-scale architecture, all paint a picture of a major settlement with at the very least, the beginnings of social complexity.

V. Early Helladic Faunal Summary

The fauna from Early Helladic Helike, Tsoungiza, Lerna and Tiryns all contribute information toward understanding the animal economy in the Early Bronze Age Peloponnese. Comparison of the taxonomic composition of each of the sites under analysis with respect to the major animal domesticates shows both distinct differences between settlements, and several similarities (*Figure 7.30*). At Tsoungiza, Tiryns and Lerna, sheep and goat are the primary animals under exploitation. This is suggestive of an economy with an emphasis on secondary products. With the exception of Tsoungiza, cattle comprise a significant portion of the faunal assemblages under analysis, while the frequency of pig is somewhat more variable. Helike is the only assemblage under consideration in which pigs account for the majority of the recovered specimens. This may reflect Helike's socio-economic position in Early Helladic Greece, and will form the basis for discussion in Chapter 8.

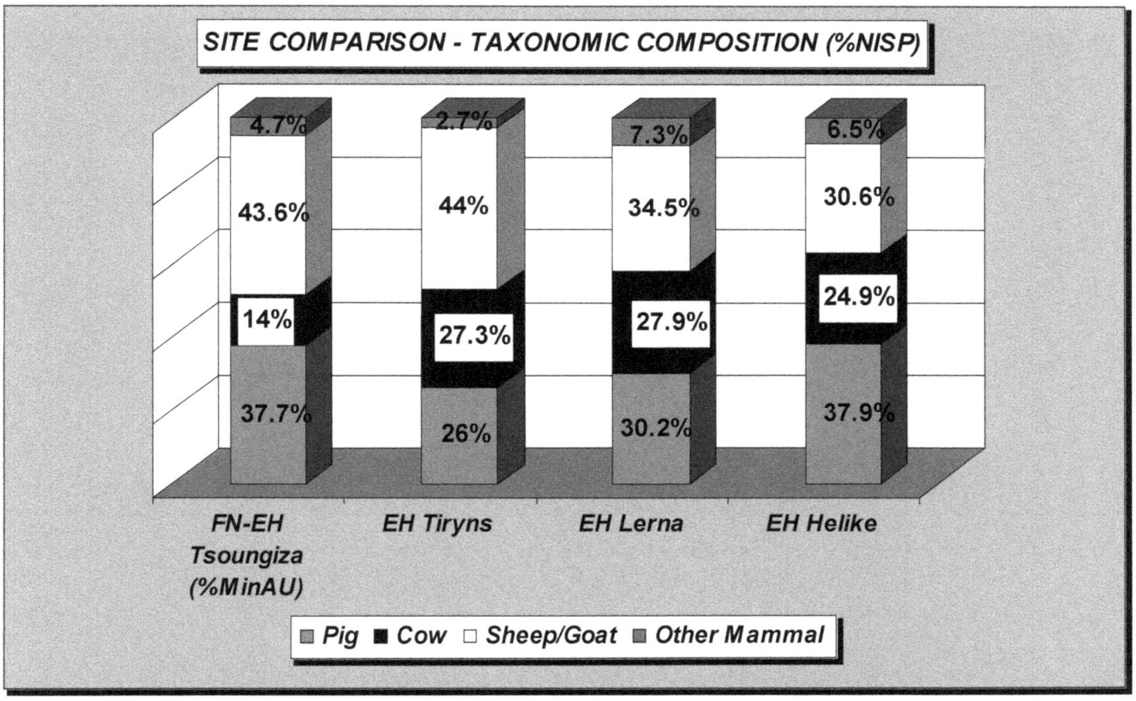

Figure 7.30 **Taxonomic Composition of EH assemblages from Tsoungiza (Halstead 2006, in press.), Tiryns (von den Driesch and Boessneck 1990), Lerna (Gejvall 1969) and Helike (Fillios 2006)**

Analysis of the relative body parts of each of the domesticates also contributes significantly toward understanding this early animal economy (*Figure 7.31*). All major skeletal parts are present, however in differing quantities. Appendicular elements form the highest percentage of those skeletal elements identified followed by cranial and foot bones.

This suggests that the major species were being bred and slaughtered on site. The high proportion of limb bones also suggests that the primary function of these animals was as meat producers. These data in concert suggest a self-sufficient animal production system. The high percentage of axial elements at Tiryns is the one notable difference, and this could be evidence that slaughter took place outside the settlement, with carcasses undergoing a small amount of transport. This would explain the relatively small distribution of cranial and foot elements.

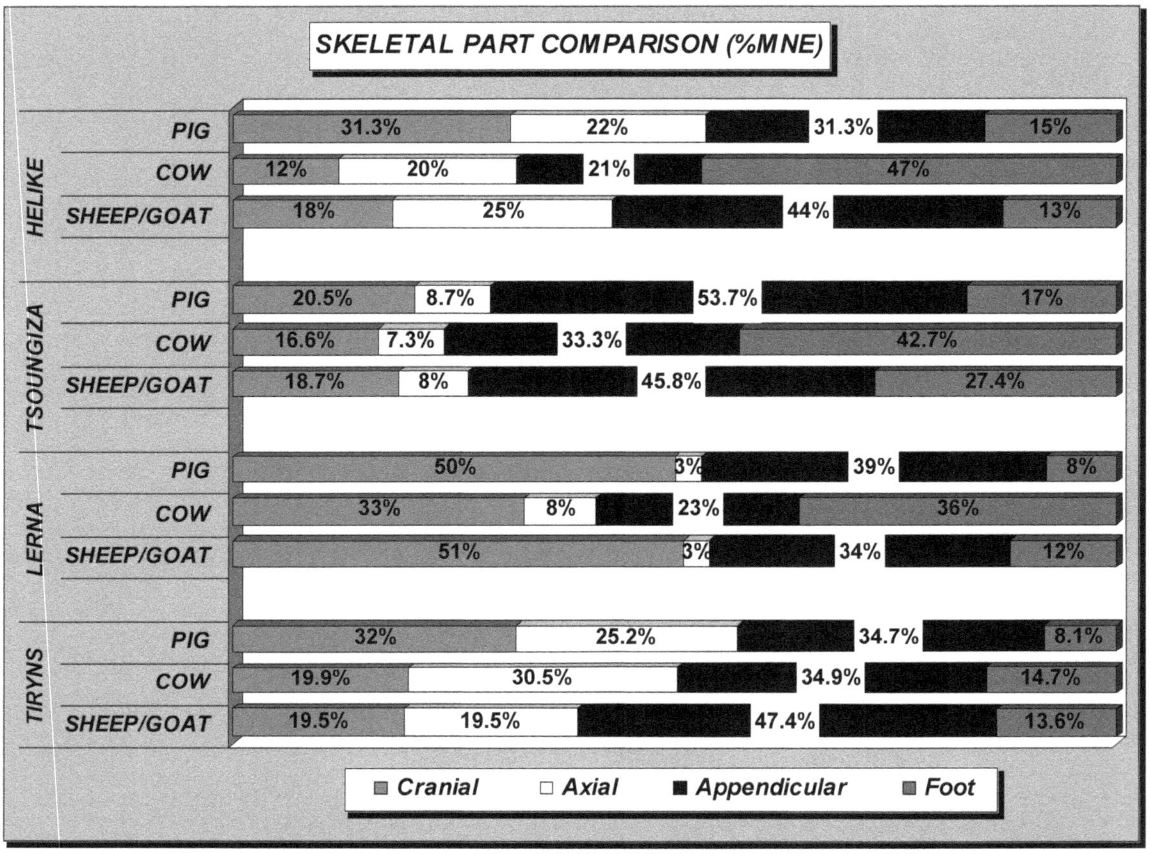

Figure 7.31 **(Lerna and Tiryns data presented in %NISP)**

In summary, the fauna paint a picture of an Early Bronze Age Greece composed of largely self sufficient settlements practicing animal husbandry – though the data also suggest that these settlements were of varying sizes, perhaps supporting Pullen's chiefdom theory. The exact nature of each of these settlements can be further refined, however, through the detailed analysis of the exploitation of one species in particular, pigs. The ensuing chapter will examine the pig data from each of these sites in detail, in order to arrive at a more nuanced image of the Early Bronze Age animal economy in Greece. This analysis will then be applied to the wider issue of social organization in this period.

PART III

ANALYSIS AND CONCLUSIONS: MAKING A SILK PURSE OUT OF A SOW'S EAR?

The third, and final, part of this study brings together all of the various aspects from the preceding chapters and creates a tableau that depicts the way in which pigs serve as an indicator of economic complexity in Early Bronze Age Greek settlements – and transitively an index of social complexity. Chapter 8 uses the general faunal data from Part II and takes an in-depth look at pigs in particular. It examines their frequency in each of the assemblages under analysis in light of several hypotheses regarding pig utilization in ancient settlements. Using the combination of the specific data with these hypotheses, Chapter 8 comes to a conclusion regarding the ways in which pigs can be used as indicators of economic systems in archaeology.

Chapter 9 looks at behavioral ecology in light of anthropological perspectives on pig utilization. It examines the ways in which biology and ecology affect and direct the human choice to manage one species over another. It addresses human-animal interactions from a general biological perspective, and illustrates how this particular perspective facilitates our understanding of the choice to manage pigs – and it's implications for human societies – both ancient and modern.

Part II of this study provided the nuts and bolts of the pig argument. It discussed the state of zooarchaeology in Greece in order to provide a base from which to view and understand the faunal assemblages from Lerna, Helike, Tiryns and Tsoungiza. Chapter 7 laid out the facts. It discussed each assemblage in order that pigs in particular be better understood within a wider context – the subject of Chapter 8.

This research suggests that pigs act as a litmus test for economic aspects of social complexity in ancient settlements. It also examines the hypothesis of the existence of social complexity in the Early Greek Bronze Age, asking the question if there is evidence for social complexity in the raising of one breed of livestock in particular. The final part of this analysis examines these issues in detail, in tandem with the suggestion that perhaps the presence of a subsistence economy is contrary to the presence of social complexity – or at the very least, is suggestive of a nascent form of social complexity.

CHAPTER 8

PIGS AS AN INDEX OF ECONOMIC COMPLEXITY IN EARLY HELLADIC GREECE

I. The Model

> *Just as the mobility and pasture requirements of sheep and goats provide nomads with a means of maintaining independence in regulated economies, the high yields and low costs of small-scale pig raising provide settled people with a similar degree of economic autonomy* (Zeder 1996:308).

This study poses the following model: Pig utilization in Early Bronze Age Greece (EH) varies in relation to important aspects of social organization, economics, and agricultural practices. The exploitation of pigs can be used to infer inter-site economic relationships, as well as the nature of intra-site subsistence economies. Pigs may serve as a "yardstick" for important socio-economic practices or changes in these practices, which ultimately aid in the reconstruction of Early Helladic social complexity and organization. Pigs are perhaps the only major domesticated species (aside from canines) that can be said to be in a complimentary rather than competitive relationship with humans (Zeder 1996:301).[50] By virtue of their omnivorous diet and their ability to find nutritional value in spoilage and waste, they are able to not only flourish on the by-products of human settlements, but they in turn provide a high protein food source back to humans. This unique relationship has yet to be examined in Greece, but has garnered attention in the Near East (Zeder 1988, 1996).

An analysis of the relationship between the intensity of pig utilization and socio-economic organization necessitates parsing out and carefully examining the various likely or plausible relationships between pig utilization and socio-economic or technological change. A significant difficulty in such an analysis, however, is the problem of equifinality: On a given site, many things may contribute to the frequency of pigs. In order to address this problem, I've assembled a series of specific hypotheses linking pigs to socioeconomic organization in Early Helladic Greece from the archaeological literature.

The following hypotheses have been used to explain and predict the intensity of pig exploitation in archaeological contexts and ancient societies. For each hypothesis, there is a culture-historical, economic, agro-technical, and behavioral biological aspect. These hypotheses variously address Early Helladic social organization as it relates to

---

[50] This complimentary relationship is situational; in areas in which intensive agriculture is practiced the relationship can be quite competitive, unless extensive measures are taken in management. This issue will be addressed in the ensuing pages.

mobility/sedentism, agricultural intensity and intra and inter site economic relationships (degree of centralization of a society), and class (social stratification). They are tested by reference to the Early Helladic faunal assemblages at Helike, Tsoungiza, Lerna and Tiryns. The bulk of this chapter will explore each hypothesis in depth with regard to the specific pig assemblages at each of these Early Helladic sites, addressing the issue of variability in the intensity of pig exploitation at the respective settlements. The end result will accomplish two significant inter-related tasks: it will create a picture of socio-economic complexity at each site, and this picture will then be used to draw conclusions about the nature of Early Helladic social organization as a whole in the Peloponnese, specifically exploring the relationship between the presence of subsistence economies in settlements and social complexity.

This chapter will conclude by comparing pig utilization among these Early Helladic assemblages in order to address the relationship between self-sustained subsistence economies and social complexity. As each of these hypotheses have been discussed in detail in Chapter 5, their explanations here will be brief.

II. Hypotheses

- *Ecological hypothesis*: Successful husbandry of pigs depends on specific geographic and climatic factors, such as the amount of rainfall in an area. As rainfall increases, pig use increases. In the absence of sufficient rainfall, substantial infrastructural investments must be made (Grigson 1987, 1995).

The ecological hypothesis is a predictive model that is perhaps best used to understand the absence of pigs in a particular assemblage, rather than an explanatory model that addresses the role of pigs in a settlement's animal economy. This model is of particular importance in the Near East, where pigs comprise just 10 percent of the exploited species in many settlement contexts (Zeder 1996). For biological reasons already discussed in Chapter 5, pigs are ill-suited to desertic and semi-desertic environments. Zeder's work at Tell Halif (1996) addressed and tested the ecological, behavioral and political aspects of this ecological hypothesis, finding that pigs could be maintained in such an environment if a greater amount of investment was made in their keeping.

In Early Helladic Greece, however, the ecological hypothesis has less relevance. Reconstructions of the Early Bronze Age climate in the Peloponnese show a similar climate to the modern day peninsula of Corinth (Corinthia). Modern average rainfall data can then be used as a proxy for Early Helladic rainfall. Grigson (1987, 1995) suggested that an average yearly rainfall between 300-350 mm was needed for minimal human investment in pig management. Figure 8.1 illustrates levels well above this (ca. 400 mm/year). Where this hypothesis is relevant is in inter site geographic comparisons. As mentioned, the issue of equifinality poses a significant hindrance to cross site comparative studies regarding the intensity of pig utilization. It is therefore necessary to ensure that all possible variables are accounted for – geographical location and climate being two of these.

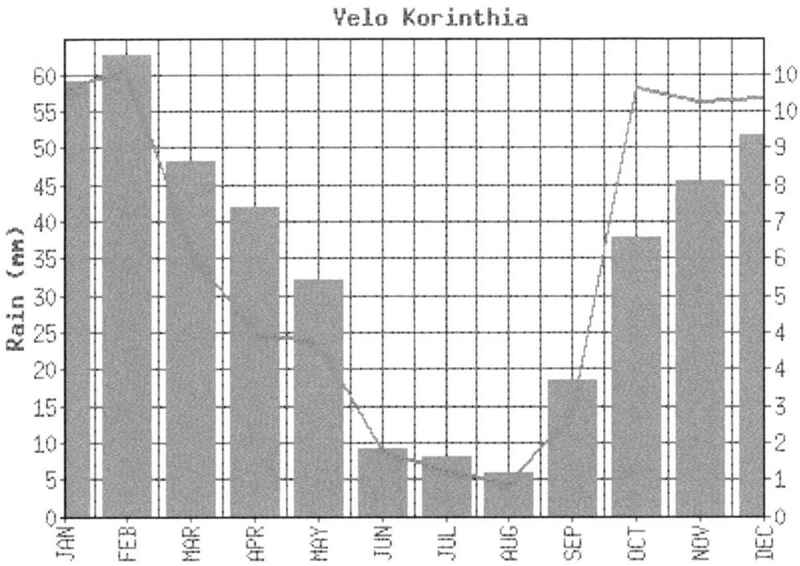

Figure 8.1 **Average monthly rainfall for modern day peninsula of Corinth**

Examination of the geographic locations of each of the sites under analysis suggests similar environmental conditions. Helike, Lerna and Tsoungiza are all located in fertile, low-lying agricultural areas with easy riverine and coastal access. Their annual temperature variations were similar, as were their rainfall levels. Tiryns is perhaps the one exception to this rule, and this is only in terms of physical location, as it sits on a hill. It still possessed access to water and had similar rainfall levels. Climate is then one factor that can be ruled out as a cause of variability in the intensity of pig utilization at the Early Helladic sites under consideration in this study. If ecological factors are not responsible for the differences in the intensity of pig exploitation between early Helladic settlements, then the question becomes, what factors were – are they social, economic, or a combination of the two? What does this tell us about Early Helladic society as a whole?

- *Mobility (transhumance) hypothesis*: The more mobile a society, the less likely it is to utilize pigs as a major resource. Pigs are difficult to herd and thus ill-suited to nomadic pastoralism. Therefore, a high frequency of pigs at a site may indicate increased or complete sedentism. Even a small number of pigs must indicate either some sedentism or links to sedentary societies (Flannery 1983; Harris 1985; Zeuner 1963).

As with the agricultural intensification hypothesis, the mobility hypothesis has less bearing on Early Bronze Age Greece, than perhaps the Early Bronze Age Near East. The existence of substantial architectural remains at all of the sites under analysis in this study already speaks to the sedentary nature of the societies. Thus, while the keeping and intensive management of pigs is a reflection of a sedentary society, this is already surmised in the Early Greek Bronze Age.

A related, more relevant issue, however, concerns ideas about private and communal land ownership. In Europe increasing agricultural intensification saw the passing of community ownership of forest and pastures to private ownership of land (Zeder 1996:302). Private land meant an end to the "management" of wild species, such as was witnessed with pigs until as late as the eighteenth century along the American frontier in the United States. This practice left pigs alone to forage. They were then harvested when needed. While seemingly unrelated to the Early Greek Bronze Age, both the American and European examples underscore important, related ideas. First, pig utilization is inversely correlated with agricultural intensification. Agricultural intensification is in turn linked to private land ownership. Agricultural intensification and private ownership of land are both components of more complex societies; in this sense pigs may act as a yardstick of social complexity. The issue of great relevance for Early Helladic Greece is the scale of the community and thus the degree of social complexity. These next hypotheses deal with just those questions.

- <u>*Agricultural intensification hypothesis*</u>: The intensification of grain production is accompanied by a shift from pig husbandry to one focused on cattle and goats (Redding 1991). Because pigs are destructive to crops there is a need to keep them out of agricultural areas. This could entail the creation of sties and more intensive husbandry/management methods.[51] Settlements undertaking intensive

---

[51] As with arid environments, pig husbandry can occur in an intensive agricultural system with increased investment and management techniques.

agriculture would reap greater benefits by breeding animals that are more compatible with this a system, such as goats and sheep. An increase in the frequency of cattle should accompany this decrease in pigs as cattle can be used as work animals for plowing necessary in agricultural intensification.

The intensification of agriculture brought about significant changes in social organization for nearly every ancient society. Storage, surplus, distribution, trade and craft specialization are just a few of the issues that arise in tandem with agricultural systems. Animal specialization and the choice to manage certain species is also a direct outcome of such a system. Because the degree of agricultural intensification is not fully known in Early Helladic Greece, animals, and thus faunal remains, become one of the best ways to approach this issue. Historical evidence provides a wealth of information on the relationship between settlements, their animals and the degree of cereal cultivation, and thus it can be used a model for prehistoric settlements. Pigs in particular can again function as a yardstick that measures the degree of agricultural production in a society, and then this can in turn address socio-economic complexity.

> Because pigs consume foods capable of supporting human populations, some have postulated that in arid areas, pigs may become a serious competitor for scarce food resources (Harris 1985). Redding (1991) has argued that although pigs are not able to process cereals as efficiently as bovids, they do raid gardens and venture into fields, consuming ripening heads of wheat and barley, trampling and uprooting plants. The invasion of agricultural fields by unpenned pigs can cause extensive damage, placing pigs in direct competition with humans. As a result…dependence on pigs…should decrease with the increased intensification of agricultural production (Zeder 1996:298).

Practical application of this theory is the best way to examine both its validity and the conclusions it suggests regarding Early Helladic social organization. Figure 8.2 illustrates the proportional frequency of pigs at each of the settlements under discussion: Helike, Tsoungiza, Lerna and Tiryns. Examination of the proportion of pigs at each site with respect to the agricultural intensification hypothesis would suggest that, since Helike and Tsoungiza have the highest percentage of pigs, they would have the least intensive agricultural systems. Conversely, Tiryns would have the most intensive system, followed by Lerna. Is this the case? Is there other evidence that supports these conclusions?

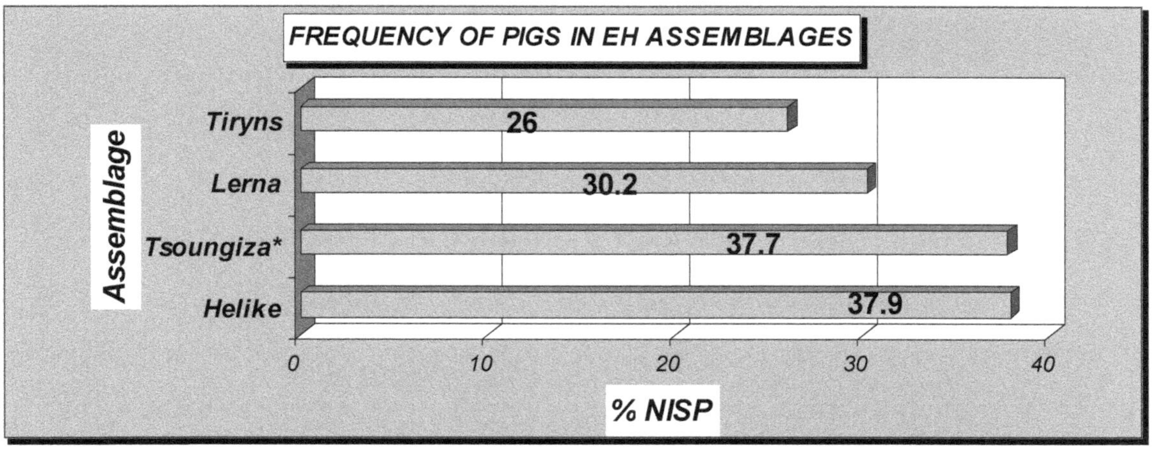

Figure 8.2 ***%MinAU (Assemblage totals (N): Tiryns N=4443, Lerna N=4738, Tsoungiza N=952, Helike N=382)**

These questions can be approached by examining the second part of Redding's (1991) hypothesis concerning the frequencies of other species in settlements. Accordingly, there should be an increase in cattle, sheep and goats. While this is difficult to see in a single chronological period, the fauna still offer information. For example, there should be a high proportion of cattle at a settlement that practices large-scale agriculture. Cattle or oxen can be used in traction, and widespread evidence exists for this practice in the ancient Mediterranean. So, if cattle were used in traction, then age profiles should show increased mortality at older ages. Their bones should also reflect heavy work with pathological markers and evidence for larger muscles at ligament attachment sites. If these types of data are available, diachronic evidence is less vital.

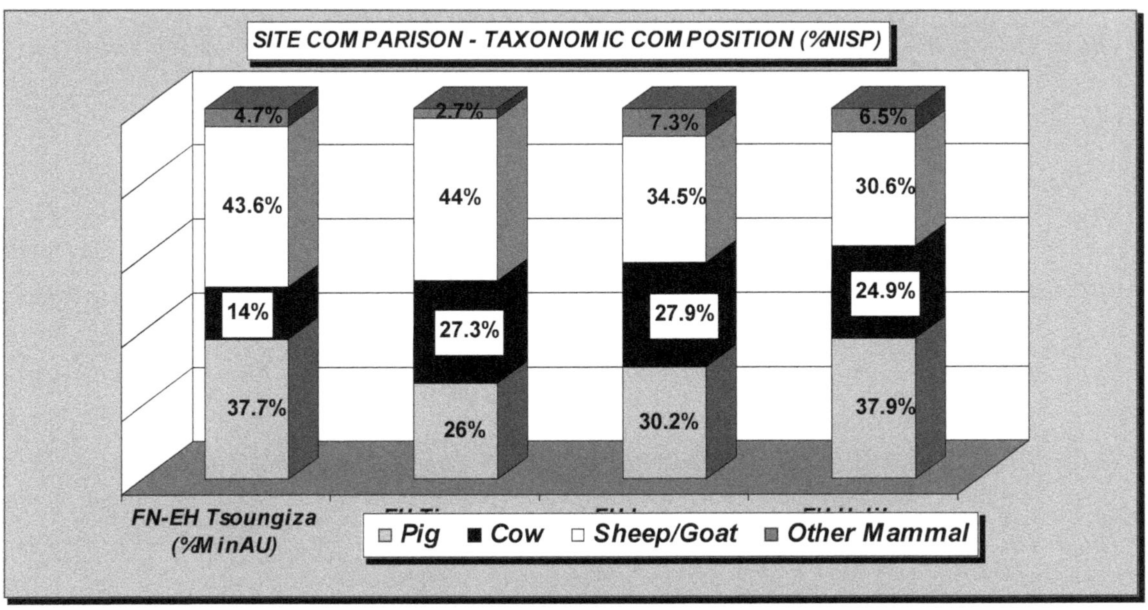

Figure 8.3 **(Assemblage totals (N): Tiryns N=4443, Lerna N=4738, Tsoungiza N=952, Helike N=382)**

Data from Lerna and Tsoungiza support these hypotheses. At Lerna, cattle comprise 27.9 percent of the total faunal assemblage (*Figure 8.3*). This is the highest percentage of all the Early Helladic assemblages under study. Gejvall (1969:32) concluded that the cattle remains from Lerna III/IV (EH II/III) represented domesticated individuals based on their small size in comparison with cattle bones from previous chronological periods at Lerna. Mortality profiles from this period show a small number of sub-adults, while nearly 80 percent of the individuals were over two years of age (*Figure 7.25*). From these, and skeletal elements representing all major parts of the animal (*Figure 7.24*), it can be surmised that cattle were being bred at the settlement. While it is possible that the older individuals were breeding stock, it is also possible they could have been used as traction animals. The faunal evidence from Lerna, coupled with the clay sealings found in the House of the Tiles, suggests a larger, administrative center. It is possible that this administrative center was also practicing agriculture that resulted in a surplus for trade. This indicates economic specialization, and thus a certain degree of socio-economic complexity.

Sheep and goat were the heaviest exploited species at Lerna, accounting for 34.5 percent of the identified species (*Figure 8.3*). As with cattle, mortality profiles indicate adult mortality, suggestive of their role in the production of secondary products (*Figure 7.25*). Over 60 percent of the ovicaprid specimens survived past their second year. Again, this emphasis on secondary products compliments additional forms of evidence at Lerna, supporting the theory of its role as an administrative center.

Returning the role of pigs at Lerna, underscores this idea. While pigs account for a significant percentage of the exploited species, they are less abundant than at Helike and Tsoungiza. Why? Examining specific data derived from the pig remains provides greater insight into this difference.

Mortality profiles generated from data derived from mandibular tooth eruption and wear stages illustrates that 64 percent of pigs at Lerna survived past their second year of life (*Figure 8.4*). An almost equal percentage of pigs were killed before the end of their first year and second year. This age profile does not reflect an economy based on intensive pig husbandry. In a system with a larger focus on breeding, a larger number of younger individuals should be expected. This abundance of younger individuals is especially predicted given the large litter sizes of pigs (8-12) in comparison with the other major domesticates (1-2).

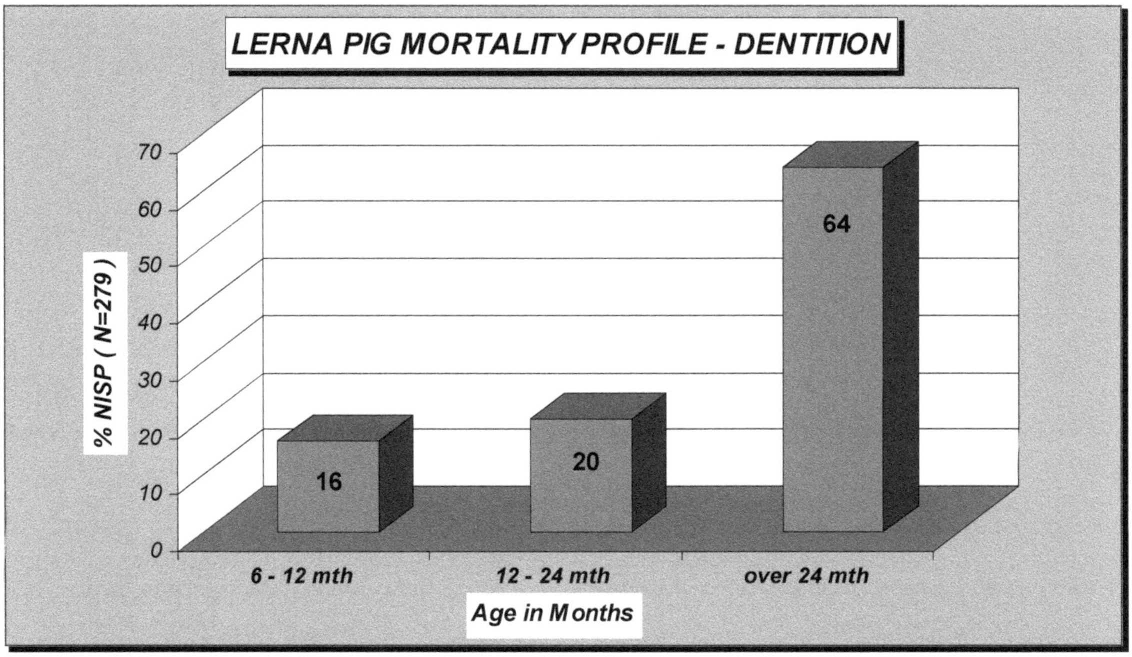

Figure 8.4 (after Gejvall 1969)

Drawing too many conclusions from mortality profiles poses certain dangers. Examination of pig mortality at Helike also suggests higher mortality rates after two years of age (*Figure 8.5*). With regard to age at death of pigs, then, Helike at Lerna look the same. What does this mean, and how does this fit in with the abundance of pigs in agricultural areas? Perhaps mortality profiles are not a good indicator of agricultural intensity and are open to too many factors to act as a good gauge, such as sample size. In modern populations of pigs, slaughter occurs between 9 to12 months of age, once an individual reaches an optimal weight. However, in ancient societies, devoid of modern, industrial breeding techniques, this optimal weight may not have been attained until the two year mark. This same mortality data shows almost no individuals over the age of three years of age. The evidence then suggests that the two year mark was the desired age for slaughter.[52]

---

[52] We can look at skeletal element frequencies to interpret carcass utilization, but a factor in this is preservation which can be affected by bone density. The densest (strongest) elements in pigs are the ulna, mandible, and distal metacarpal. Patterns of skeletal element representation are obscured when fused/unfused elements are pooled. A general implication of this finding is that age at slaughter can seriously bias the calculation of mortality profiles and species proportions since not all species are slaughtered at the same age and fused/unfused bones have different rates and modes of survival (Ioannidou, 2003:364). Ioannidou's study shows that there is significant inter-taxonomic variability in bone density values which means that different species do not survive equally, however there is less intra-taxonomic variability in density for pigs. In pigs, females may actually have denser bones than males, but the data were conflicting. Sex differences for pig were influenced by age with females seeming to have higher density when adult.

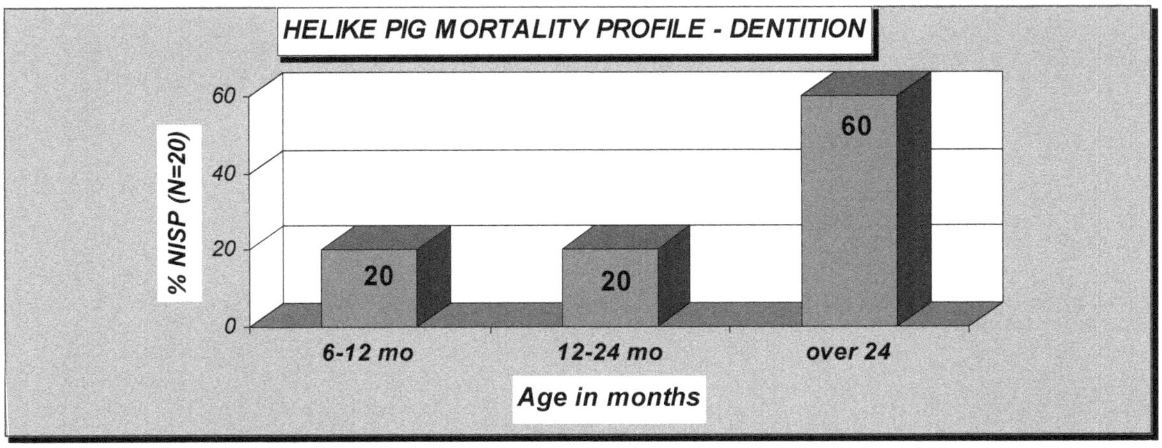

Figure 8.5

Mortality profiles from Tsoungiza further complicate this picture. At Tsoungiza it appears that the opposite phenomenon occurred. Young individuals are quite common and older individuals less so (*Figure 8.6*). Halstead (in press) concluded that most deaths occurred before or on the threshold of adulthood between the first and second to third years. He believes this pattern is consistent with both the high reproductive rate of the species and the absence of their role in the production of secondary products. He also suggests that the existence of all ages of pigs may be a by-product of slaughtering for immediate consumption, rather than the modern practice of preservation. Maybe the problem lies with being overly concerned with differences that may not be significant in the end?

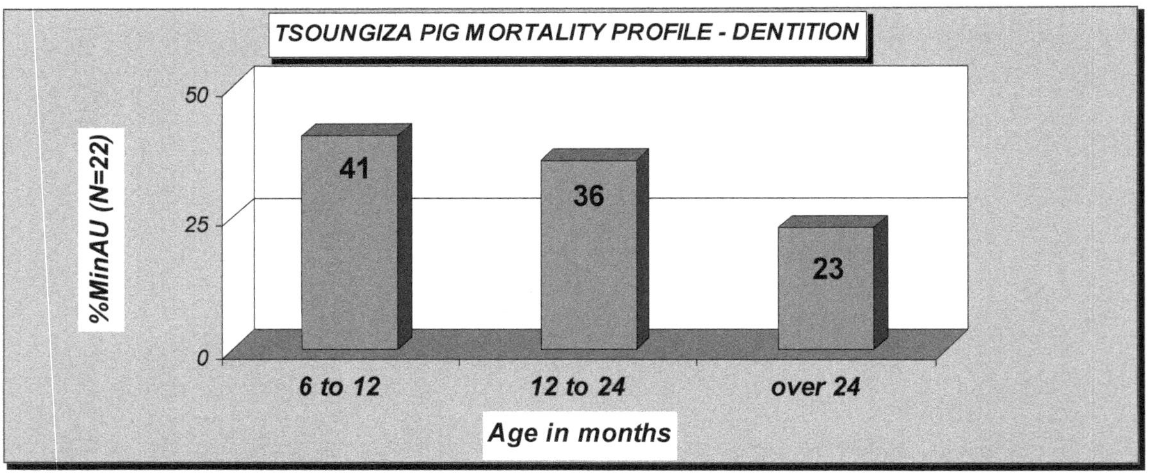

Figure 8.6 **(after Halstead, in press)**

Mortality profiles from Tiryns reflect the opposite pattern from Tsoungiza. At Early Helladic Tiryns the majority of pigs appear to have been slaughtered over 24 months of age (*Figure 8.7*). However, as with Tsoungiza, younger individuals are also present. If all individuals up to two years of age are viewed together and then contrasted with older pigs, then the majority of pigs may have been slaughtered in their prime. In this case, the presence of older individuals in the assemblage suggests some probable breeding stock, mixed, perhaps, with stock obtained from the surrounding area.

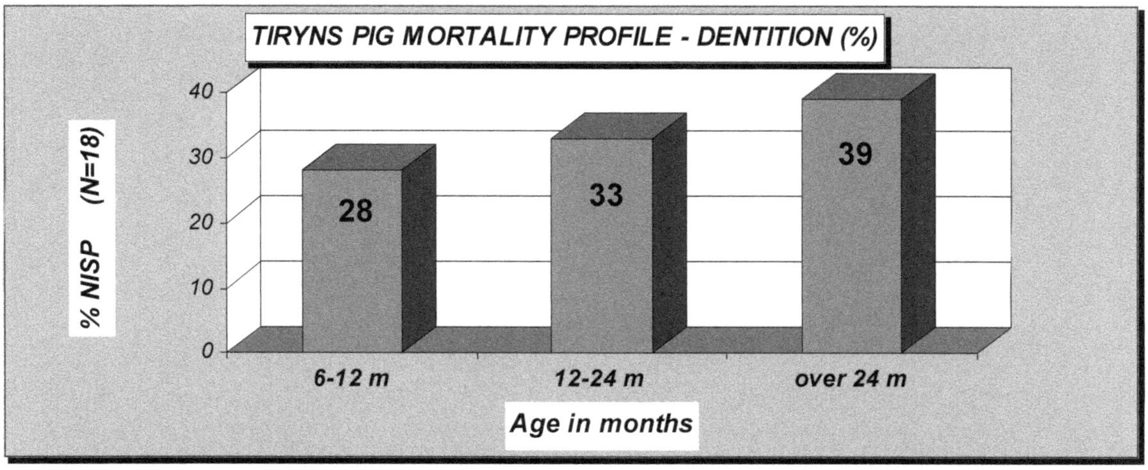

Figure 8.7 **(after von den Driesch and Boessneck 1990)**

Comparison of the mortality profiles between Lerna, Helike, Tsoungiza and Tiryns may reflect smaller differences than originally thought – and perhaps their relevance to the agricultural intensification hypothesis is also greater than the data initially suggested. Figure 8.8 illustrates an interesting pattern of pig mortality: Helike, Lerna and Tiryns mirror one another more closely, while Tsoungiza reflects the exact opposite pattern. This suggests that Helike and Lerna may be more similar to one another than either is to Tsoungiza. But intensity of pig exploitation at Tsoungiza and Helike closely resemble one another, while Lerna is most similar to Tiryns. Thus, maybe the best way to utilize mortality data is by looking at the mean age at death of the individuals, rather than examining the actual frequencies of death at each age. If the mean age at death is calculated for each assemblage, the data is much closer to Halstead's for Tsoungiza, and an average slaughter age becomes somewhere in the one to two year age range. This is consistent with local rearing practices, and supports a local pig economy at each site. Adhering too strictly to age determinations based on subjective criteria in a variable species for which there is no hard and fast reliable data on tooth eruption, may, in the end, create false problems. Perhaps the data needs to be used more generally. Then slaughter profiles become less of a problematic area and relative frequencies of the pigs at each site becomes more important, more of an indicator.

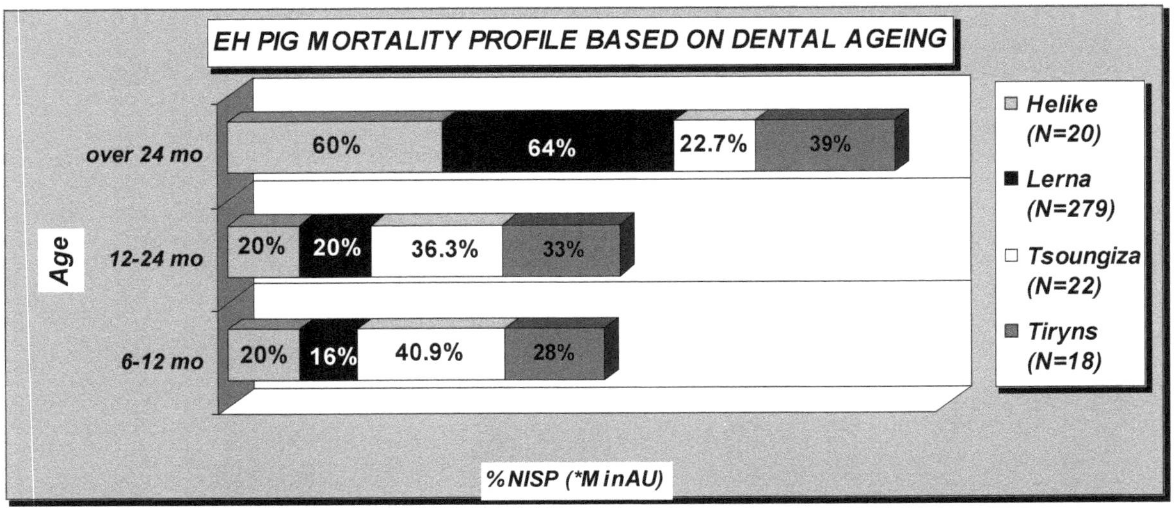

Figure 8.8

Examining mortality profiles suggests several conclusions with respect to theories about the relationship of pigs and the intensification of agriculture. It suggests that mortality profiles are a useful indicator as to the place pigs were reared. The existence of all age classes illustrates that pig husbandry is occurring locally on site. At Lerna, Tsoungiza, Helike and Tiryns, it appears that pigs do not play a role in provisioning – that is, pigs are not an item that is being brought into these settlements in exchange for another commodity. The data suggest that while mortality profiles are useful, they may not be directly useful as an indicator of agricultural intensity. Hypotheses need to be based on other sources of evidence – such as relative frequency of individual species.

According to species representation, the Early Bronze Age settlement at Tiryns should have practiced the largest scale agriculture of all the sites under consideration. Is this true? While only a small portion of the Early Bronze Age settlement at Tiryns was excavated, more recent re-excavation of the large circular building (Rundbau) brought to light new evidence that suggests the function of the building was as a granary (Killian 1986). If accurate, the structure's function strongly suggests that the settlement may have practiced intensive agriculture. Conclusions suggested by the presence of a granary can be combined with those suggested by the frequency of pigs. If Redding (1991) is correct, and the frequency of pigs is inversely proportional to the scale of agriculture in a settlement, then Tiryns must have possessed a larger scale of agricultural production than both Helike and Tsoungiza. Since marked differences exist between the assemblages with respect to the frequency of pigs, there must a host of factors responsible. It is highly possible that agriculture is one of these factors.

A related issue in this discussion with direct bearing on the faunal assemblages is the scale of agriculture at the settlements under analysis, and this is a contentious issue in Early Bronze Age Greece. At Tsoungiza there is figurine evidence clearly illustrating the knowledge of yoking pairs of cattle (Pullen 1992). But this does not mean that Tsoungiza practiced large-scale intensive agriculture. Halstead concluded that the faunal assemblage from Tsoungiza exhibits several features consistent with small-scale mixed farming, and few features representative of specialized, extensive herding (Halstead, in press). Are the faunal assemblages from Helike and Tsoungiza representative of small households practicing garden cultivation? Perhaps Helike is a small settlement, with small scale cultivation being practiced by households/ or a groups of households? What could this look like archaeologically? Pigs would likely be a common feature, and they could be fenced out of these cultivated areas. These questions bring up additional questions surrounding the size of the settlement and its' inter site socio-economic ties and regional position.

> While agricultural intensification may be a factor in shifting pig use, it would not appear to be the only factor. Agricultural intensification…need not have resulted in the elimination of pig from the diet. Instead of abandoning pigs as an important dietary resource, the conflict between crops and pigs could have been eased through a shift in management techniques from extensive foraging to sty-based pig management is unlikely with the level of technology available in ancient times. Successful pig management during periods of agricultural intensification would have to take place on a small-scale, household level basis (Zeder 1996:308).

Thus the real issue that pigs may address is not the presence or absence of intensive agriculture – but the scale as it pertains to the household and the settlement. All of the assemblages under analysis in this study suggest local and consumption of pigs. This suggests relatively self-contained systems of animal husbandry, and by extension, autonomous subsistence practices. However, these practices may be at the household level in smaller settlements, and at the community level in larger settlements. In either case, the data strongly supports the notion that the intensity of pig exploitation is inversely correlated with the intensity of agriculture. Pigs can therefore function as a proxy indicator of economic complexity with respect to agriculture in ancient settlements.

- *Centralized society/Urban economic hypothesis (Political economy):* The frequency of pigs should be inversely correlated to the degree of centralization of a society (Diener and Robkin 1978). As settlements become differentiated in size and function, variance in the intensity of pig use should increase between sites.

Intensive pig husbandry is a good strategy for small, independent settlements because pigs yield a high amount of meat and can be easily fenced out of small, garden cultivation areas. In a smaller settlement in which intensive agriculture is not practiced, utilizing pigs as a prime meat source would require minimum investment, and may help enable a community to maintain independence from larger, 'administrative' centers or centers of authority. Pig husbandry is a useful rural subsistence strategy that permits satellite communities to emphasize their domestic, non-market based modes of production and, in so doing, maintains a degree of independence from those centers in the political economy which seek to control them (Diner and Robkin 1978).

Theories concerning the intensity of pig use and centralized economies are closely linked with the intensity of agriculture in a society. With centralized economies and hence economic specialization, come social complexity and socio-economic differentiation. As these things occur, Diener and Robkin (1978) predict that the intensity of pig husbandry would decrease *at some sites.* If this is the case, then it is most beneficial to compare and contrast the relative frequency of pigs between different settlements (*Figure 8.9*). In addition, if a settlement spans several chronological phases, the frequency of pig intra-settlement can be compared in order to ascertain whether there is a change over time and nature of the settlement. This approach touches on the idea that pigs may be an initial subsistence strategy and fall in disuse as a site matures into a larger center.

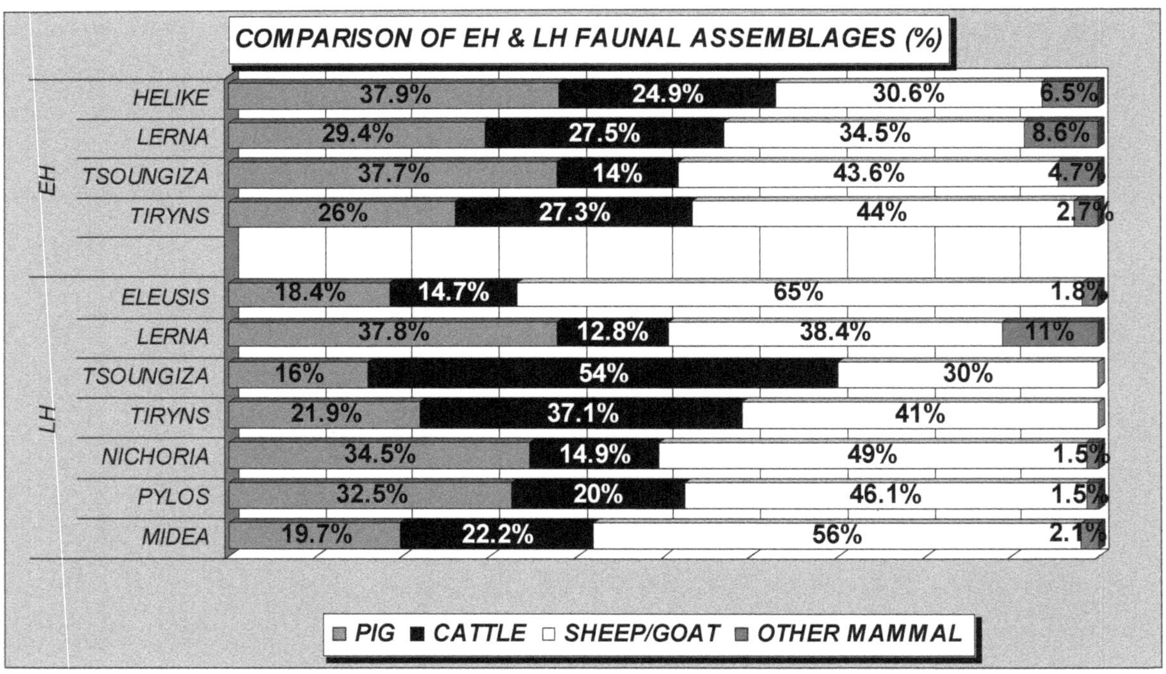

Figure 8.9

As with any data set, questions and analyses are limited by the corpus of material. The sites under discussion here lend themselves best to inter site analyses of frequency and intensity of pig exploitation. By approaching the assemblages within this framework, an image of possible centralization in Early Helladic Greece is created.

Examining the frequency of pigs at Lerna, Tsoungiza, Helike and Tiryns from the perspective of this centralized society hypothesis suggests that Helike and Tsoungiza would be the most autonomous settlements (highest frequency of pigs), while Lerna and Tiryns (lowest frequency of pigs) would be the most integrated into a wider social network (*Figure 8.2*). Here it is necessary to look at other forms of archaeological evidence to test whether these predictions could be correct. Certainly at Lerna there is substantial evidence for contact, and most likely trade, with surrounding settlements (Chapter 2). The clay seals and imported pottery both attest to Lerna's contact with other settlements in the region, and probably beyond. Evidence from Tiryns, most specifically the presence of the Rundbau structure and its probable function as a granary, certainly suggests a wider regional role. If both Lerna and Tiryns were regional centers on a larger scale than other settlements, functioning as administrative centers, then according to this hypothesis, they should have relied less intensely on pigs – and they did in comparison to Tsoungiza and Helike.

What about evidence for Tsoungiza's regional role? While a few cylinder seals were discovered at Tsoungiza, its role in relationship to other Early Helladic settlements is still under analysis (Pullen 2006, in press). According to Halstead, the faunal assemblage at Tsoungiza is representative of small-scale mixed farming. Mortality profiles for all domesticates indicate the presence of all ages of animals, and skeletal element frequencies of the major domesticates indicate that all the parts of the animals were present, especially pigs (*Figure 8.10*). These data support Halstead's conclusion that Tsoungiza had a self-contained animal economy. He also suggests that while knowledge of using oxen for plowing was present, this does not indicate large-scale agriculture. Returning to the idea of the inverse relationship of pigs and centralized societies and applying it to Tsoungiza suggests that Tsoungiza is not part of a wider hierarchical settlement network. This does not mean that Tsoungiza did not trade with other settlements, *but it does suggest that there may not have been a large degree of economic integration within a larger Early Helladic settlement system. Does this suggest that an overarching socio-economic hierarchy did not exist?*

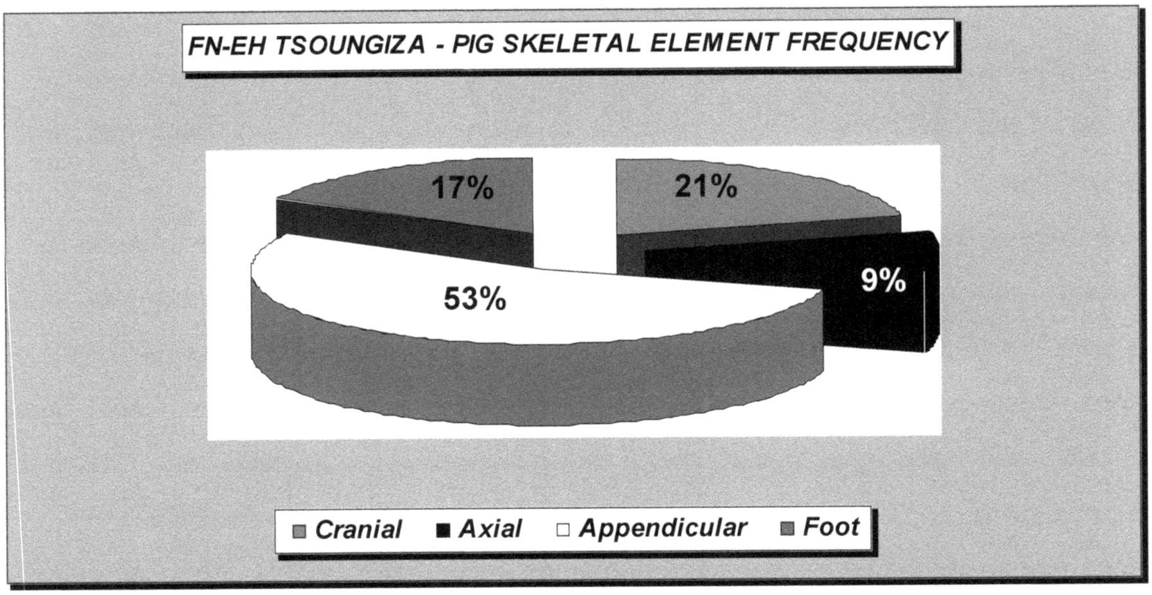

Figure 8.10 **(expressed as %MinAU N=359)**

The faunal evidence from Helike also offers some insight into this question. Helike has the highest frequency of pigs of any of the settlements under examination. If this frequency is taken as representative of the actual frequency, and not an artifact of a small sample size or contextual bias, then Helike should have been an independent settlement outside of the socio-economic control of a larger settlement, such as perhaps Lerna. In many respects, the same conclusions drawn for Tsoungiza can be drawn for Helike. With regard to subsistence, Helike most likely possessed a mixture of small scale mixed farming coupled with mixed animal husbandry. The mixed composition of livestock at Helike with sheep, goat, cattle and pigs all well represented offered a wide range of animal products that could be exploited, and this mix could act as a buffer against a range of ecological or environmental factors minimizing risks to livestock. This mix of animals would also have made use of a wide range of local sources of food and browse in the keeping of the livestock, a system in which pigs would have been valued for their flexibility. The existence of mixed livestock also provides evidence of the scale of the settlement at Helike. Mixed livestock with varying demands of food, water and shelter are difficult to mange in large numbers – again consistent with small-scale farming, and most likely a small-scale settlement. These conclusions are all supported by data derived from mortality profiles and skeletal element frequencies.

One significant difference exists with respect to Helike that is not present in the other assemblages if mortality based on epiphyseal fusion is examined (*Figure 7.7*). Individuals aged 12 to 24 months are conspicuously absent when age at death is determined by this method. If this is an accurate reflection of pig age profiles, than there is the distinct

\

possibility that Helike may be provisioning pigs to other settlements. A possible role as a provisioning site might help to explain the high proportion of pigs at Helike. The question that this possibility brings forth is what is Helike getting in exchange? At this stage, this is a question that can only be answered by further excavation.

> I would argue, then, that the reason pig use declines is not due to the degree of agricultural intensification, but rather to the degree of integration into regional economy that most often accompanies periods of agricultural intensification...Since pigs do not lend themselves to centralized production in non-mechanized societies, but are most successfully raised by individual consumers, swine would not be a favored resource in complex, often centrally coordinated urban economies...(Zeder 1996:308).

One last hypothesis need be briefly mentioned before all the data can be pulled together. The class hypothesis perhaps has less relevance to the Early Bronze Age than it would for Medieval Europe, but its examination underscores some important issues.

- *Class hypothesis*: The production and consumption of pork is associated with individuals of lower or working class status (R.T. Hecker 1982; Panagiotakopulu 1999). Pigs reproduce quickly, are omnivorous, and provide a cheap and relatively efficient way to feed workers. Additionally, small households could afford to keep several pigs as food sources.

The relationship of pigs with social class extends far beyond economic issues and into the realm of taboos and lore that surround modern day pork consumption. Marvin Harris and others have addressed these issues in great depth, and so they will not be reiterated here. What does need to be highlighted, however, is the suitability of pigs to smaller households of limited means. For reasons already discussed throughout the body of this work, pigs are the most economical and practical animal to keep as a food source of all the major domesticates. They should therefore play a large role in ancient societies with a smaller degree of social complexity. The faunal evidence from Helike, Lerna, Tiryns and Tsoungiza all support this idea. Is the conclusion, then, that Early Helladic Greece did not have a great degree of socio-economic complexity?

PART III

SOCIO-ECONOMIC COMPLEXITY IN EARLY HELLADIC GREECE? THE CONCLUSION

> ...pigs become a gauge of social and economic cohesion, a barometer of individual autonomy and independence, a bell weather of social change that allows us to examine more fruitfully some of the basic forces that shaped the peoples of the ancient Near East, their economy, their politics and their beliefs (Zeder 1996:309).

Although Zeder's work focused on pig exploitation at Tell Halif in the Near East, it is just as applicable to Early Bronze Age Greece. The examination of pigs in relationship to each of the hypothesis examined above suggests that the intensity of pig exploitation is related to a host of external issues, and in the absence of other forms of data, can be used as an indicator of socio-economic systems in ancient societies. Therefore, pig data from all four Early Helladic sites suggest several conclusions:

- There was variability among settlements in Early Helladic Greece, both in settlement size and in degree of socio-economic inter site relationships. Helike and Tsoungiza were most likely smaller, fairly autonomous settlements, while Lerna and Tiryns were more likely to have played wider regional roles.
- Pigs can be used to indirectly reconstruct social complexity via their role in economic organization in ancient settlements. Thus, while Helike may have been a self-sufficient settlement with regard to food production, provisioning of other, larger settlements may have occurred. This would provide an economic or trade relationship linking Helike to surrounding settlements. However, the data do not suggest a wider socio-political relationship with surrounding settlements. Daniel Pullen's chiefdom hypothesis is supported by the frequency of pig utilization in regard to differentiation in settlement size. If pigs are a valid indicator of settlement size, then Helike and Tsoungiza were the smallest settlements, and Tiryns and Lerna the largest.
- Early Helladic Greece is a period marked by the beginnings of social complexity with regard to the creation of new economic relationships between sites. The frequency and intensity of pig exploitation taken in conjunction with analyses of other animal domesticates provides an understanding of these relationships.

Zeder writes that "pigs are less common or absent in periods of greater integration within a regional urban economy" (1996:307). In order to test her conclusions, and those just posed by this research, there is need for diachronic studies. While all of the sites here do not support such a study, general comparison can be made with other periods in Greece for which socio-economic complexity is better known. If all the proposed conclusions are indeed tenable, then the frequency of pigs in faunal assemblages should decrease through time. Pigs should be more prevalent in the Early Bronze Age, a period marked by the formation of new socio-economic systems, and least prevalent in the Late Bronze Age (at some sites), a time of marked social stratification. This is indeed the case, as illustrated by Figure 8.11.[53]

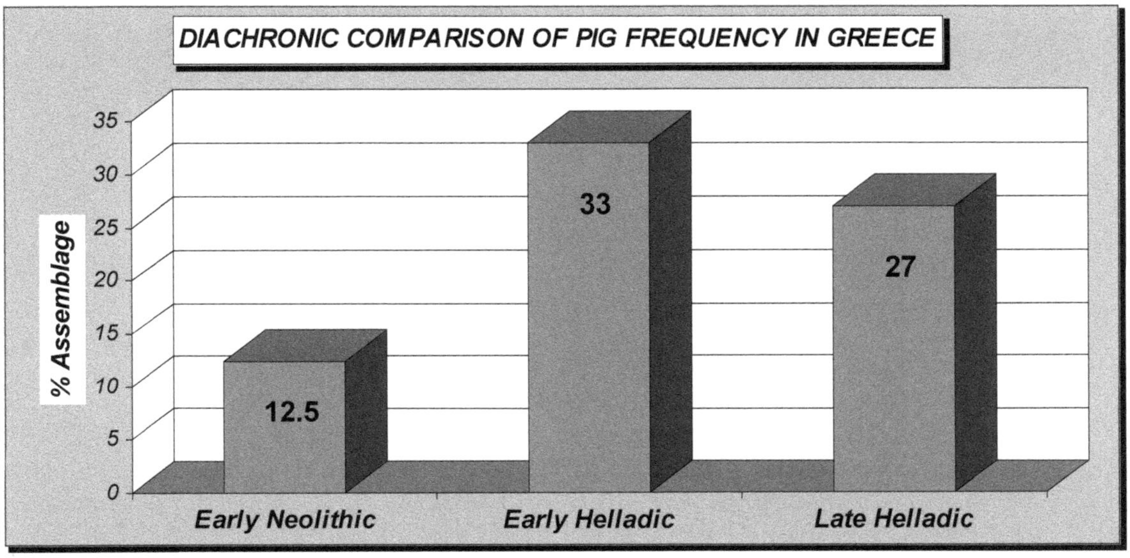

Figure 8.11 **(Summary Neolithic data after Halstead 1992).**

This chapter has shown that the frequency of pigs in ancient settlements is a barometer for economic complexity, especially in regard to the intensity or scale of agriculture. At those sites which are thought to have more "complex" administrative and economic systems in place, such as Tiryns and Lerna, the frequency of pigs is lower than in those settlements lacking these systems, such as Helike and Tsoungiza. Pigs are then barometers of economic organization and can be used as proxy indicators of *economic* aspects of social complexity. The data and conclusions supported and suggested by pigs paves the way for an exploration into the notion that a self sufficient animal economy may run counter to the existence of social complexity in societies. Pigs provide a stable, steady

[53] While evidence from the Neolithic would seem to run counter to these conclusions, it must be remembered that domestication began during this time period, and for many sites the species represented may have been wild.

subsistence source, allowing households or even small villages, to maintain their autonomy – at least in regard to integration within a larger regional economy. Since integration within a larger, regional economy often implies, or at the very least, leads to political/administrative control by the larger settlements, a high proportion of pigs suggests self-sufficient settlements both politically and economically. If economic factors are used as proxy indicators for socially complex societies, than we may be able to surmise that indeed the presence of autonomous animal economies suggest a less socially complex society – or in the case of Early Helladic Greece, a society that has yet to attain the complexity of later periods.

The next chapter continues in this vein with a discussion of behavioral ecology. Until this point, this study has made mention of this body of theory, without delving too far into its merits. Chapter 9 presents a critical discussion of the utility of behavioral ecology as it pertains to human-animal interactions. This discussion will illustrate the necessity of accounting for biological and ecological factors in any and all discussions of animal management by humans. It will illustrate how biological factors are truly behind the "choice" to manage one species over another. This biologically driven choice in turn shapes and governs the socio-economics of animal management.

CHAPTER 9

GENERATION OF SWINE: EVOLUTIONARY ANTHROPOLOGICAL PERSPECTIVES ON PIG DOMESTICATION IN BRONZE AGE GREECE

Throughout this analysis I've attempted to incorporate, where appropriate, principles of Behavioral Ecology of human-animal interactions to understand social complexity in Bronze Age Greece. The present chapter represents an explicit effort to consider the domestication and use of *Sus* (as compared with other domesticates) in this framework. What biological factors render an animal a good choice for exploitation; what are the social and economic effects that harnessing certain animals may have on a group of people; and what are the effects of domestication on local human ecology, economy, and social organization? More particularly, why were pigs so heavily exploited during the Greek Bronze Age (see Chapter 5) and why has there been an apparent decline in the use of this animal in recent centuries?

In taking this approach, it is appropriate to broaden the analysis to consider domesticates in general (in a phylogenetic framework); to divide the animals into relevant "morphs" (within species) such as gender or level of maturation; to list and quantify key biological variables; and to consider the potential mutualistic, parasitic, and competitive interactions between potential domesticates and *Homo sapiens*.

I. Phylogenetic Framework

A biological analysis of animal use strategies must take into account the fundamental biological differences among possible domesticates, including variables ranging from body size and growth rate to sociality and mating systems. Some of these details will be dealt with below. Initially, it is worthwhile to consider the phylogenetic relationships among these animals, because this reflects and encompasses all of the key biological factors.

Table 9.1 represents a simplified structure of the relationships between potential domesticates in the Mediterranean/European region. Sheep and goat are hard to distinguish in archaeological faunal assemblages for good reason: they are closely related and overlap a great deal in their adaptations, including body size and mode of locomotion. Sheep, goat and cattle (referring here to all domesticated species) are members of a diverse clade (*Bovidae*) which in turn is a member of the very diverse order Artiodactyla, also including pigs, antelope and deer.

ORDER	*FAMILY*	*SUB-FAMILY*	*SPECIES (Common Name)*
Perisodactyla	Equidae		Horse
Artiodactyla	Suidae	Suinae	Pig
	Bovidae	Caprinae	Sheep
	Bovidae	Caprinae	Goat
	Bovidae	Bovinae	Cow
	Bovidae	Antilopinae	Gazelle
	Cervidae	Cervinae	Deer
Carnivora	Canidae	Canis lupus	Dog

Table 9.1: **Taxonomic relationships between major domesticates**

Within the *Bovidae* family, diverse species such as sheep, goat, cattle, deer and antelopes form a clade that excludes the suids, and correspondingly, share a number of key biological traits having to do with reproduction, locomotion, feeding ecology and mating system, that are different from those of *suids*. All of these – sheep, goat, cattle, antelopes, and deer – are candidates for domestication in that they are common and would comprise the majority of the wild medium to large mammal meat supply for any foraging group in the circum Mediterranean region. Why, then, have only some been domesticated?

Globally, there is evidence for management of deer in various places, but with the exception of recent developments in deer farming,[54] nothing like domestication is documented. It has been argued that gazelle (subfamily *antilopinae*) were either heavily managed or domesticated in Neolithic West Asia (Moore 1978). Nonetheless, only sheep, goat and cattle among the *Bovidae*, were domesticated to any important degree.

This observed pattern in the history of domestication - which species underwent domestication and which did not in this diverse group - may have an explanation rooted in the biological differences among them. If such an explanation exists and can be unambiguously stated, then the traits identified as either making some species amenable to domestication, or traits identified as making some species unlikely to be domesticated, define a sort of biological criteria set for this type of human-animal interaction. It is beyond the scope of this study to explore this issue in detail, but it is most likely that body size, reproductive patterns, or maturation rates do not explain this pattern. These variables are either very similar across these species, or are highly variable within each group (especially body size), so that they cannot cleanly divide these species into "haves" and "have-nots" for the "right stuff" for domesticability.

[54] The farming of red deer (especially in New Zealand) is a recent innovation that has a large body of literature (see in particular Drew, 1976).

The most likely variables that potentiate or limit domestication are related to sociality and mating systems. It is most likely that true herding behavior, consisting of female bonded groups monopolized by one or few males, selects for behavioral traits that allow humans to intervene and control feeding patterns, movement or transhumance, and breeding. Animals that have either small "family groups" or that breed in leks are less likely to be controllable in this manner.

This type of difference in grouping and mating behavior is said to be the explanation for the inability of humans to domesticate zebras, while at the same time being able to domesticate horses. These two closely related and in many ways similar species have dramatically different social systems, the former grouped in families (which are in turn grouped in sometimes very large "pseudoherds") while the latter form true herds with stallions (Diamond 1997).

This phylogenetic analysis indicates that there are certain species of *bovids* that may be considered a closely related group that is qualitatively similar in many key variables. Putting this more bluntly, it could be said that at one level, sheep, goats and cattle are all different versions of the same kind of animal. Much like Homer Simpson's magic pig, this "domesticatable" group offers a great deal of variety in body size, and thus time to achieve maturity, and some variability in secondary products, allowing a wide range of pastoral strategies. But in all cases, certain key biological variables such as reproductive rate, diet, and absolute growth rate remain roughly the same, or almost identical when scaled to body size (see next section).

The pig, on the other hand, is different. As will be discussed below, *Sus* is not merely just another domesticated edible mammal. Indeed, pigs are nearly as different (or more different?) from sheep/goat/cattle as are dogs, despite being in the same Linnaean order. When comparing among domesticates, it is not biologically justified to consider sheep, goat, cattle and pig as equally or interchangeably different. They are all different from one another, but some (the pig, as Orwell foreshadowed) are more different than others.

II. Morphs as Units of Analysis

It is now appropriate to turn from the most general (the broad phylogenetic context) to the very specific (morphs within species). William Hennig, the founder of cladistic theory, was probably the first to systematically incorporate the fundamental and obvious fact that differences between morphs within a species must be taken into account when considering traits that distinguish species.[55] Otherwise a butterfly would be considered very distantly related (phylogenetically) to the caterpillar from which it metamorphosed. Here, it is important to consider the different morphologies of domesticated animals not to perfect phylogenetic taxonomy as was Hennig's goal, but rather to properly isolate and formulate socioeconomic questions that must have been important in animal husbandry and the economic and social realms of animal use.

Consider this question: What is gained, and what is lost, by the slaughter and consumption of a particular beast? What is gained is always a certain amount of meat, and in some cases, terminal secondary products (such as bone, tooth/tusk, and hide). What is lost depends as much or more on the morph as it does on the species. If the specific animal is a rare breeding bull, then future reproductive activities are terminated. If the specific animal is a young male pig, then little is lost, but if the specific animal is milk producing cow, then what is lost is a lot of future milk. It is reasonable to divide lost opportunities into four categories: 1) future additional meat (a non-zero quantity in a growing animal); 2) future renewable secondary products such as milk, draft power or fleece; 3) future use in breeding (may apply to very few individuals); and 4) future use in special social contexts such as ritual or (possibly) marriage related price or payment.[56]

The importance of morphology is underscored when it intersects with phylogenetic considerations. With respect to costs and benefits of husbandry choices, as well as the broader question of impetus to domesticate or manage a certain species, the sheep/goat/cattle group appears even more different from pigs, given these considerations, than the more traditionally figured biological variables indicate. For instance, if the practice of keeping pigs involves breeding semi-feral sows with fully wild boars, a roasted male piglet never represents a lost breeding opportunity. Since there is no evidence of breeding cows, ewes, or nannies with wild bulls, rams, or billies, husbandry of sheep/goat/cattle will frequently require consideration of future breeding value.

---

[55] Hennig, W. 1965 *Phylogenetic Systematics*, University of Illinois Press. Urbana.

[56] Utility in marriage related price or payment may simply be the same as the animal's economic value as measured from cost-benefit analysis, but it may be wise to leave open the possibility that an animal used in this manner has additional sociopolitical or even ritualistic value.

III. Key Biological Variables

There are numerous biological factors relevant to the choice to domesticate or manage, as well as numerous methods for doing so for *bovids* (cattle, sheep, goats) and *suids* (pigs), and there are also many ways to describe or classify these factors. Following is one possible list of factors to consider.

- Life history parameters
    - Size at birth
    - Maturation rate
    - Age at first reproduction
    - Litter size
    - Interbirth interval
    - Length of reproductive career
    - Total lifespan

- Somatic variables
    - Body size (by sex and age)
    - Daily caloric demand under relevant conditions (of age, gender, lactational state, etc.)
    - Type of food required (e.g. high vs. low quality)

- Adaptive features that result in secondary products
    - Hide qualities
    - Fur or fleece
    - Horn or tusk

From these can be derived a number of variables more directly linked to management and domestication, including:

- Total production per unit area of space
- Total production per unit time or resource invested
- Specifics of management strategies

- Appropriateness or viability of domestication vs. management of wild population
- Differing role in local economy linked to settlement size
- Direct or indirect competition for resources (the fodder may be human food)
- Exchange value of the animals in relation to the local food economy
- Exchange value of the animals in the broader social context

Biogeography as a metavariable:

Biogeography is a broad factor encompassing environmental variables such as listed above as they are manifest in the context of climate, topography, access to water, and by extension, access to food (graze, fodder). In the Mediterranean region, the main animal domesticates at most archaeological sites are cattle, sheep, goats and pigs. Since these animals were consistently present, the first question that comes to mind is why? At a broad level, their existence in the human faunal record (of foragers and farmers) since nearly the first human settlements in the area can be explained by examining the geographical distribution of their wild forms. Figure 9.1 is a reconstruction of the geographic areas in which wild cattle, sheep, goat and pigs were likely found during the early Holocene. These areas, namely a large portion of the Near East and the Mediterranean, correspond roughly to the areas in which they are found in archaeological assemblages. This correspondence between wild and domesticated animal distribution suggests that people initially exploited the animal resources that were locally available in their respective settlement areas. If geography were the only variable that early societies took into account in deciding which animal to exploit, then the indigenous distribution of wild species would suggest that most human settlements should have relied heavily on pigs and cattle, as they were more widespread than sheep and goats (Figure 9.1). (Not to mention the other medium to large body size mammals that were not domesticated anywhere). Faunal analysis reveals that this is not always the case. This is in large part due to the myriad of factors that contributes to the decision to exploit one species over another. In addition to explanations offered by behavioral ecology, issues of trade, management and the complex issue of domestication all come into play in these decisions (Chapter five discusses the issue of management and domestication in greater depth). It is sufficient to point out, based on the distribution of wild species indicated by Figure 9.1, that some degree of trade or cultural diffusion of ideas played a role in the spread of these animals to other areas as part of the process of "domestication."

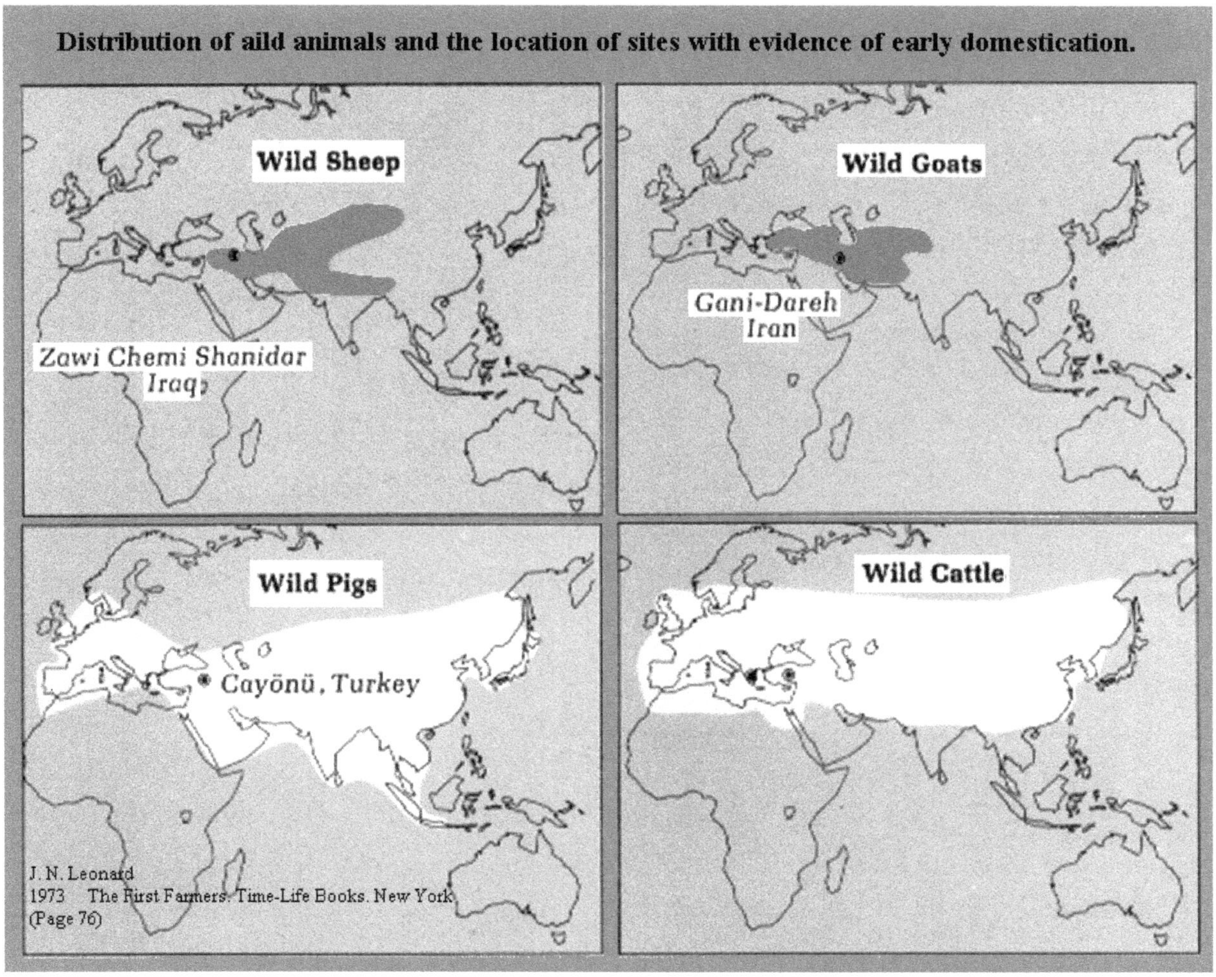

Figure 9.1

Examining the natural habitat of each of these species provides insight into animal management decisions. Each animal's unique biological traits are integral into its eventual viability in a geographic area. If an animal is unsuited biologically to a certain environment, a greater degree of human intervention and effort would be required for their successful exploitation. The following discussion will address the specific environmental niches to which each of the major species is best suited with an eye toward adding to explanations of their relative abundances in different geographic areas.

Capra:

Wild goats (often called *Capra aegagrus*) had the smallest geographic range of any of the future animal domesticates (Figure 9.1). Largely confined to the Near East, they eventually spread (by human agency) to Africa, the European Mediterranean, and most parts of the globe. Their widespread distribution suggests that while they prefer certain environmental conditions, they are quite adaptable. However, given the difficulty in distinguishing between sheep and goats osteologically, their natural range may have been larger – closer to what is indicated for sheep. Faunal evidence suggests that goats were the earliest ruminant to be domesticated (Zeuner 1963:129). Although sheep and goats are more often than not grouped together during faunal analysis, they do differ, primarily in their ecological requirements and in the raw materials they supply. Goats are browsers, preferring to feed on the foliage of shrubs over grass. They are well adapted to higher elevations above the tree line in mountains, and in lower, arid zones (where small shrubs are abundant) (Zeuner 1963:130). Their ability to also feed on herbs unsuitable for other ruminants allows goats to thrive further into the desert than most domesticates. While both sheep and goats are known for the secondary products they offer, goats produce more milk than sheep, probably being exploited for dairying before cattle.

Ovis:

The utilization of sheep (*Ovis aries*) by humans followed goats closely in timing, and in some regions it has been argued that it was the first animal to be managed (Zeuner 1963:154). The natural range of wild sheep is similar to goats; however they appear to have extended their range further east into Asia. As with goats, however, they soon became widespread through human agency. However, due to a host of factors, the environmental range of sheep is more limited than goats – most varieties do not fair well in hot, desert climates (an exception being the fat-tailed sheep of the Kalahari). Their fleeces render them ill-suited to some of the African and Near Eastern desert areas in which goats flourish, as well as the humid tropics.

The fodder requirements of sheep also differ from goats. While goats are browsers, sheep are grazers, preferring a diet of grass. Sheep also prefer more protected environments than goats, although this preference is not a large factor in regard to the actual management of sheep. In comparison with goats, sheep provide less milk, but do provide a slightly higher quality meat (*Table 9.1*), in addition to the wool harvested from their fleeces.

Thus with respect to environmental and climatic factors, behavioral ecology predicts that goats would be preferable to sheep for heavy exploitation in Greece. The majority of Greece is rocky, relatively dry, mountainous country, more suitable to goats than sheep (if ease of exploitation is the sole factor). What behavioral biology can not predict is the human factor – those secondary products that are desired, whether for subsistence or economic reasons. Thus there is an underlying human element that must always be considered alongside this evolutionary perspective. This human element coupled with an evolutionary and ecological perspective offers a good explanation, however, as to why sheep and goats are the heaviest exploited species in Greece. It also provides a good rationale for a second look at settlements in which they are not the heaviest exploited species (such as Helike).

Bos:

Cattle (*Bos taurus*) provide a good illustration of the interplay between human desires and environmental and biological factors. Based on the distribution of wild cattle *(Figure 9.1)*, it would seem that they should be ubiquitous in ancient settlements, and present in high numbers. In reality, what this distribution illustrates is their suitability to the environment found across the Mediterranean, Europe, Asia and North Africa. The behavioral ecology of cattle alone can not account for their presence/absence or frequency in faunal, assemblages, but a deeper understanding of their life history variables can. Unlike the other ruminants (sheep and goat), cattle require considerable attention and organization – extensive and intensive human effort. It has been suggested that intensive human utilization of cattle in Europe did not fully come about until the establishment of agricultural practices (Zeuner 1963:241) (though of course, this would not apply in Africa where extensive reliance on cattle predates agriculture in many regions). Although evidence for cattle predates intensive agriculture in European archaeological settlements, the link between cattle and agriculture may well be strong.[57]

Cattle are grazers, like sheep, (though quite capable of eating a wider range of herbaceous vegetation and browse as a fallback food) however they require large, enclosed pastures with plenty of grass. This necessitates an environment capable of supporting their needs, and while they are indigenous to Greece, they are an animal that would necessitate a certain degree of mobility in order to obtain their required fodder. This study would go a step further and link the intensity of cattle exploitation in Greece to the private ownership of land (also tied to agricultural intensification), as a necessity in cattle husbandry. In societies that practice intensive cereal agriculture, of the type certainly in place by the Mycenaean Period, private ownership of the land under cultivation is generally in place (though it may be owned by the palace). As cattle are an animal that require large areas to graze, they become impractical animals in areas without land on which to keep them. Thus, they will become relegated to those places wealthy enough to possess the space (and subsequently, fodder)

---

[57] Cattle pastoralism and the "cattle culture" not only predates agriculture in many areas of Africa (Gifford-Gonzales 2005). More importantly, the sequence of agricultural developments involving both plants and animals, the nature of domestication, and the relationship between social structure and food production is very different in Saharan and Sub-Saharan Africa compared to Mediterranean Europe. This difference is so large that it is far beyond the scope of this chapter and study.

necessary in their keep. This is an issue well-worth exploring in future research, but one which must unfortunately be sidelined in this study.

The exploitation of cattle, then, works with the products AND the services the animal renders. Cattle produce milk, however goats were used for this far earlier, in greater numbers, and are far easier to keep. Cattle also supply strong hides and a large quantity of meat. However, their utility must be seen as economic. Though their products play a role in their exploitation, the amount of human effort and time required to bring one cow to slaughter far exceeds the effort required for other species. Their greatest distinction from the other *bovids* is in their use as beasts of burden, especially in traction. In addition, their dung provides a good and abundant source of fuel in areas where wood is scarce, and it can also be used as fertilizer in agricultural areas. But this discussion of cattle brings up the related issue of human utility – or what the benefits reaped are for the work put in.

Sus:

While meat per se is not the sole reason animals were exploited in ancient times, it does play an important role in the choice to exploit one animal over another. This point is especially dramatic with pigs. The distribution of wild boar is nearly as great as cattle, although it did not initially extend into climates quite as hot as cattle (*Figure 9.1*). As already discussed, pigs are the most flexible of the major domesticates in regard to diet. Like humans, they are omnivorous, capable of living on nearly anything but preferring mast, roots, green plant matter – and agricultural crops. Their diet also contains seeds, nuts, fruit, vertebrates and invertebrates (Shley and Roper 2003). The wide variability in the diet of wild pigs suggests they are opportunistic omnivores, whose diet in any particular instance is strongly influenced by availability. This adaptability is one of the driving factors in their widespread use and domestication diachronically. A likely downside of the versatility of pig diet is that it is also a more demanding one in terms of quality. A vast grassland with little forest or brush will not support *suids* but will support *bovids*.

The environment of pigs is most delineated by climate. Although adaptable, they do require adequate amounts of shade and water for thermoregulation, preferring a habitat containing forested areas. Study of the density of wild pig populations suggests that access to riverine woodlands is the determining factor in group size and habitat selection (Choquenot and Ruscoe 2003; Virgos 2002). Specifically, research suggests that proximity to riverine woodlands constrains foraging, since riverine woodlands contain better quality pasture than the surrounding shrublands. Preference for more sheltered habitats may impart other benefits as well, such as refuge from predation or hunting and ready access to water.

Pigs do not offer renewable secondary products like the other major species exploited by humans, and so in their case, behavioral ecology offers an explanation as to their utility. A look at the amount of meat produced by all the animals under consideration highlights this point (*Table 9.2*). Pork offers the highest caloric content of all the major domesticates and the most fat. Based on their nutritional content, keeping pigs is good subsistence strategy.

IV. Biology of Choice and Methods of Domestication and Management

Table 9.2 provides summary data comparing goats, sheep, cattle and pig, including life history and productivity variables. Also included are a few illustrative calculations using these data. The pattern that emerges from these data support the assertion made above that while all these species are different, pigs are more different. Goat and sheep are small bovids, and cattle are large bovids, and while there is probably no overlap between *ovicaprids* and *bos* (indeed, there is a gap in body size and all of the concomitant variables), it could be said that in many ways *ovicaprids* are scaled down cattle (or cattle are scaled up *ovicaprids*). The most startling finding of this analysis (though also admittedly somewhat whimsical) is this: Starting with a single breeding female, and keeping and raising all individuals to maturity for ten years, cattle, sheep or goat would yield a mere 12 to 24 thousand kilos of fat (perhaps equal to the fat in two years' supply of ice cream for the Greater Boston Metropolitan area, personal observation), while pigs would yield over one trillion kilos of fat, or approximately 500 times the total mass of humanity alive on the planet earth today. Thus, in terms of medium to long term strategy, the choice to raise pigs vs. cattle is not a trivial one, if fat is important.

When examining the relative benefit of managing a certain species one must examine what the species in question can offer via subsistence, and conversely, what it will necessitate to care for and sustain until the point of slaughter. Table 9.2 compares each of the major exploited species with respect to their particular life history variables. Of course, these variables are inextricable from environmental constraints, but nonetheless they warrant their own detailed attention. Also, the table does not include information on ranges or variability, because that is beyond the scope of the present analysis.

VARIABLE	CATTLE	SHEEP	GOAT	PIG
Phylogenetic Group (as used in this analysis)	Bovidae	Bovidae	Bovidae	Suidae
Average Lifespan (yrs)	20	10	15	17.5
Average Number of Births per Year	1	2	2	15
Average Number of Births per Lifespan	10	20	10	262.5
Type of Food Required	Grass/Browse	Grass/Browse	Browse/Grass	Omnivorous
Quality of Food Required	Low Quality	Low Quality	Low Quality	**High Quality**
Average Body Weight	1200	120	100	250
Average Body Weight (Kilos)	462	46	38	96
Number of months to reach maturity	12	12	12	9
Number of animals from a breeding female in 10 years (with continual breeding, first breeding at one year, and if all were kept)	512	19,683	19,683	68,719,476,736
Growth Rate (Kilos per month)	15.3846154	2.307692	1.923077	10.68376068
Average Meat Yield (lbs)	456	46	35	113
Average Meat Yield (kilos)	175	18	13	43
Kilos of Meat Per Month over Lifespan	6	1	1	5
Calories per Kilo	2,020	1,490	1,450	3,710
Protein per Kilo (grams)	190	170	160	140
Fat per Kilo (grams)	140	90	90	350
Total Calories per animal	354,277	26,362	19,519	161,242
Total protein per animal (grams)	33,323	3,008	2,154	6,085
Total fat per animals (grams)	24,554	1,592	1,212	15,212
Calories per month over lifespan of animal (calories from meat divided over months of maturation) if killed at maturity	11,809	1,318	976	**17,916**
Protein (grams) per month over lifespan of animal	1,111	150	108	676
Fat (grams) per month over lifespan of animal	818	80	61	**1,690**

VARIABLE	CATTLE	SHEEP	GOAT	PIG
Total amount of meat in a hypothetical 10-year reproductive spree (see above, this table)	233,472	905,418	688,905	7,765,300,871, 168
Total amount of fat in a hypothetical 10-year reproductive spree (see above, this table)	12,571,569	31,341,392	23,846,712	1,045,328,963, 426,460
Renewable Secondary Products	Milk (potentially large quantities) and dung (potentially large quantities)	Wool (potentially large quantities), some dung	Some hair (rarely, only in some varieties) and some dung	Some dung. Bristles may be used but typically are taken during butchery
Terminal Secondary Products	Hide (probably of higher utility than the other animals under consideration), bone, horn	Hide, bone, some horn	Hide, bone, some horn	Hide (but probably of lower utility than cattle/sheep/go at, bone, tusk
Possible Morphs (in ancient context)	Bull (breeding male); steer and oxen (castrated male) dairy cow; young individuals for slaughter	Adult males and females, young	Adult males and females, young	Adult females and young, adult males in systems where they are kept for breeding, otherwise not.

Table 9.2 **Selected life history and productivity variables for cattle, sheep, goat and pigs. The bolded items are of special interest (see text).**

The respective biological traits of each of the animals under consideration have a direct impact on their economic utility to humans. These traits play a critical role in understanding the resulting human-animal interactions that shape societies. If, for example, the average lifespan is examined for each animal, it is clear that cattle, having the longest lives, will take the longest time to reach maturity. From a subsistence standpoint, then, cattle would not be a wise choice to manage if regular access to food from meat was the main concern.[58] Since they average one birth per year, and only ten births per lifespan (assuming they live to old age), their management requires a large expenditure for comparatively smaller return than the other species. The large body size also means that cattle require more fodder than the other animals, thus requiring greater resources. Although one cow yields more meat than any of the other animals, this can be more of a hindrance than anything else, if the cow is being utilized by just one family. However, if cattle offer other benefits, such as milk, hide, traction, then the heavy human expenditure their care necessitates is offset by these other, economic, factors. If the meat from one cattle is shared or traded, this further

[58] Large scale production of cattle would provide this need. If the faunal assemblage of a region had cattle as its only domesticate, one could reasonably suspect this. Again, however, this analysis is focused mainly on Mediterranean Europe and West Asia. Were Africa considered, it would be necessary to add blood as a renewable secondary product for cattle, and of course it turns out that there are regions in Africa where cattle comprise the sole domesticate.

increases the economic utility of the animal. Thus the life history variables associated with cattle provide insight into the specific conditions under which heavy exploitation is most likely to occur - this should be in larger settlements that practice more intensive agriculture.

Sheep and goat have similar biological traits vis à vis reproduction, body size and average meat yield. Thus they should be the preferred choice for exploitation in smaller settlements as they reach sexual maturity at an earlier age, and require less food and less investment in their keep. They will have more births during their life spans (which are shorter than cattle), and these factors, coupled with the larger amount of secondary products they offer, have rendered them a ubiquitous resource in ancient settlements of all sizes.

Pigs, however, are perhaps the best resource to manage for individual households or for smaller settlements, for several reasons. Pigs are distinguished from other domesticated animals in their high number of offspring and short time to sexual maturity. A sow averages eight piglets per litter, with some capable of bearing up to 12. If contrasted to the singleton births of sheep, goat and cattle, it is easy to understand why pigs appear so frequently in faunal assemblages. In addition to bearing a large number of young, a sow can reach sexual maturity during the period from 7 to 22 months, while males reach sexual maturity by their tenth month (Dobney 2000; Tikhonov et al. 2004). And though wild boars usually bear one litter per year, pigs are capable of bearing two, as modern sows are in heat for a period every three to four weeks. They have easier birthing than the other animals, and due to the high number of offspring, relatively higher rates of juvenile survivability. This prolificy, coupled with a flexible diet, and the fact that they require little in the way of specialized care and offer a relatively high meat yield when their total life spans and litter sizes are taken into account, are all biological traits that render them an asset to humans and a wise choice for exploitation.

Though pigs provide no major secondary products (hide and tusks aside), the large quantity of meat could be used to an individual family's economic benefit by providing a commodity that can be traded for other goods. Pigs are still small enough that when slaughtered, they provide abundant meat, but not the excess of cattle. As already pointed out, the main hindrance to raising pigs is their destructiveness to agricultural crops – and this factor affects their abundance in archaeological assemblages.

V. Hogs vs. Herds: "Generation" of Swine as a Unique Strategy

Can the differences in the economic significance of harnessing pigs as a resource, compared to sheep, goats and cattle, be explained by reference to biological variables such as body size, rate of reproduction, and dietary regime? Since *bovids* (cattle, sheep, goat) and *suids* have different environmental and subsistence needs, how do these ecological requirements affect the decision to manage one "domesticate" versus the other?

The above analysis, admittedly preliminary, suggests these and similar questions can be addressed and understood with a behavioral ecological perspective. While the biogeographical framework for wild forms of these domesticates guides us to the relevant variables, it does not explain the pattern of selection among husbandry strategies (in particular, which animals to raise) seen in the faunal record. This could be simply because humans are always good at buffering their livestock from the environment, and certainly this is true to at least some extent. However, at a global scale, it is simply not true that humans raise whatever animals they want wherever they want to, and to the extent that this has occurred more frequently in recent centuries, it is done so with special (newer) varieties, technology for exploiting subterranean water sources, etc.

However, certain variables do seem to explain what the faunal studies suggest to be an inordinate fondness for pigs in small-scale insipient state societies. These are primarily reproductive rate, body size and time to maturity, fat content, and secondary products. With respect to these variables, there are certain likely choices among the bovids, with the preferred strategy being to avoid cattle except under certain circumstances. But among the four main domesticates focused on here, it is clear that pigs are, as in the words of the spider, under certain circumstances "terrific."

The uniqueness of the pig as a domesticate means that the pig is a proxy indicator for social and economic structure. It has been possible, in this analysis and in Chapter 5, to define the conditions under which the pig *is* terrific and to contrast this with conditions under which it is not. The pig, though its bones, speaks to us (*senu* Binford) about the nature of Early Bronze Age Greek society, and the changes that were occurring at that time.

VI. Conclusion

Debate surrounding the archaeological identification of domestication has already been addressed. However, behavioral ecology's guiding role behind the human exploitation of animals must be understood in order to underscore the necessity of incorporating this framework into faunal analyses. As already pointed out, changes in animals resulting from domestication can be genetic, behavioral, and a combination of the two. These changes in turn have a tremendous effect on the social and economic systems within a society. Biological definitions of domestication focus on genetic changes in animals as a result of selective and artificial breeding. Socio-cultural definitions focus on the relationship between humans and animals – both however are inter-related. Thus analyses of human animal interaction – the core of animal exploitation – must be studied from both a biological (genetics, behavioral ecology) and socio-cultural perspective.

Approaching questions surrounding domestication from both a biological and socio-cultural perspective makes clear the need to heavily apply principles of behavioral ecology to the data. The application of behavioral ecology to the study of pigs in Bronze Age Greece can reveal otherwise unexplored aspects of human-animal interaction – a key organizing feature of every human society. This study has used proxy indicators in the faunal record to measure or assess:

- the relationship between climate change and the distribution of pigs;
- the relative abundance of domesticated bovids, pigs, and human settlements in different natural and human created habitats;
- the relative abundance of pigs in relation to indicators of the socio-economic complexity of a settlement; and
- the potential correlations between the type of settlement, i.e. administrative center versus small village, and the frequency of pigs.

Applying the concepts of behavioral biology to the study of animal exploitation enables the examination of the life-history parameters of each of the major species to be compared to one another in order that animal husbandry and exploitation may be viewed as a dynamic human-animal interaction, in which each constrains and changes the other in turn. The following questions can then be asked of the fauna: How do certain animals, such as pigs, fit into this approach? How can a behavioral ecology framework address the socio-economic organization of small-scale societies? Do faunal remains suggest a use other than food?[59] If the fauna does suggest subsistence, what parts of the animals are commonly represented, and what can skeletal element frequency indicate about utilization or social status? What do certain animals suggest about the ancient environment?

Ultimately, domestication involves altering the natural history/life history/ biology of an animal in a way that improves some product or effect (selective breeding) desirable to humans. Domestication may well have occurred in degrees, however, and so incipient domestication could have involved simply altering one or two traits, while "full" domestication involves bringing an animal to the point where some degree of dependence on human care emerges (even if only partial dependence).

Complicating matters further is management. Management is a term strictly applied to wild animals where their numbers, distribution, sex ratio, etc. are controlled typically through the process of selective culling or similar activities. Management differs from hunting in that in hunting one kills what one finds, while in management, harvesting selected individuals, or harvesting at particular times (i.e., avoiding harvesting during pregnancy and lactation, or avoid harvesting females in general, etc.) is forgone. Some degree of control and choice is exercised and selective culling practiced.

Hunting with even a modest bias (such as a cultural norm that says the hunter can't eat the meat of a pregnant animal, thus causing hunters to occasionally not shoot the pregnant animal) is management, but maybe just a little management. Even a small amount of management can have huge effects, however. And - not all domestication rules make sense in either the emic and/or etic sphere. At the other end of the spectrum, intensive management may have strong effects on a population, to the extent that it fits part of the definition of domestication. Perhaps strategic flux is the term that would work best for the regular movement between these strategies.

Returning to the actual analysis of the fauna in archaeological contexts highlights the problems with a domestic/wild dichotomy. As clearly shown, there are a myriad of factors that contribute to the choice to harness one animal over another, and in turn, a myriad of variables act on individual animals – the culmination of their life histories, how they were used, their diets, and many more. Faunal analysts pride themselves in their ability to analyze

[59] Recent research at the Late Bronze Age sanctuary of Ayios Konstantinos at Methana in the eastern Peloponnese yields abundant evidence for pig sacrifice (Hamilakis and Konsolaki 2004). Do pigs serve a similar function at other sites?

these bones and draw conclusions as to the individual animal in question's life. But what about biological factors that can't be accounted for, or readily quantified? What about allelic variation, norm of reaction, or simply observer error? There are so many factors that shape a bone, that to draw a line between wild and domestic on the basis of bone morphology is to potentially make molehills into mountains. An understanding of the role of evolutionary anthropology, biology and ecology and its place in every individual's life, illustrates the inherent problems with a focus on "domestication".

Behavioral ecology is an integral factor in human-animal interactions. Examining faunal assemblages from within its framework adds a missing dimension to analysis of the original animal population that aids in understanding how animals were used by ancient populations. Behavioral ecology is a concept that must be employed in the final analysis of any animal population, before conclusions are drawn as to particular species' probable relationships to humans. Though largely a "common sense" framework, it is rarely acknowledged as a contributing factor in the human choice to manage one species over another. Biological and ecological factors are integral components behind economic decisions concerning animal management, and these in turn are key factors in reconstructions of socio-economic complexity in Early Bronze Age Greece.

CHAPTER 10

CONCLUSION: ODE TO THE PIG. EARLY HELLADIC SOCIAL COMPLEXITY & THE PIG LITMUS TEST

The faunal assemblages from Helike, Lerna, Tsoungiza, and Tiryns form a corpus of data that provides substantial information about human occupation in the Peloponnese during the Early Bronze Age. In particular, the frequency of pigs in these settlements acts as a barometer of social complexity, especially as it pertains to economic factors. The analysis of the proportion of pigs in relation to cattle, sheep and goats provides a wealth of information on ancient subsistence patterns, exchange networks, and economic specialization. In particular this study concludes:

- The frequency of pigs is a proxy indicator for economic organization in Early Helladic settlements, and transitively has direct bearing on social complexity
- There is a marked difference in the frequency of pigs between sites that is not due to environmental factors, and therefore must be indicative of inter-site socio-economic variability
- Pigs were most likely reared locally, suggesting the presence of a local subsistence economy at all four sites under study; the presence of a subsistence economy appears to run counter to high social complexity, suggesting that Early Helladic Greece is in the naissance of a socially complex society.

It was the original intent of this study to utilize the frequency of pigs in Early Helladic faunal assemblages to address all aspects of social complexity–political, social and economic. However, regardless of the nature of the original question, in the end, the data tells its own story and suggests those conclusions that are possible. In this light, the original intent must be reformulated. Specifically, the analysis of the fauna suggests that pigs are best used – in this period in Greece – as economic indicators. Their utility as economic indices does not obviate their role in social complexity. Rather, it underscores that their frequency in faunal assemblages in this region best addresses those aspects of social complexity that are economic in nature.

As Chapter three has shown, social complexity is comprised of a variety of interwoven components. The inseparability of the political, economic and social variables that contribute to complex societies has direct bearing on these conclusions. It suggests that pigs do indeed address social complexity – but they do it transitively through economic avenues. Thus, the relatively low frequency of pigs at Tiryns, in comparison with Helike, Tsoungiza and Lerna, suggests that Tiryns had a more centralized economic system with a focus on intensive agriculture. This economic specialization at Tiryns is further reflected in, and supported by, more recent excavation of the Rundbau, which is now interpreted to have functioned as a granary (Kilian 1986). Surplus goods, economic specialization, and the related issues these variables suggest, such as a more centralized political system coupled with social stratification – are all predicted by the low frequency of pigs at Tiryns. In this round about way, pigs are a proxy barometer of social complexity. Thus faunal analysis facilitates the creation of a model of the animal economy, which then facilitates the production of a model of the larger economy. Taken in conjunction with other forms of archaeological evidence, such as monumental architecture and variability in the size of structures, a fuller picture of the organization of the site emerges. This research has shown, however, that this picture can begin with pigs – and this is especially important for those sites that yield fewer forms of evidence.

Many archaeological sites do not yield enough evidence to support an analysis such as Tiryns. Although the Early Bronze Age phase of the site has not been intensively excavated in comparison with the later Mycenaean phase, Tiryns has yielded a variety of different types of evidence, including monumental architecture of "known" function. It can therefore be used as a model illustrating how pigs act as these barometers of social complexity.

Social complexity is comprised of craft/economic specialization, political organization, social stratification and urbanization. All of its respective components are inter-related, with any one of these factors having the potential to shed light on the other. Little is known of many of these aspects in Early Helladic Greece. Excavated settlements are few, and published reports even fewer. No one site affords an opportunity to view political, economic and social variables of the type about which archaeologists theorize and write. However, various types of data from different settlements are available.

Some settlements may provide evidence for material culture that is best used in analyses of social stratification (such as human burials). Other settlements provide evidence of centralized administrative/political function (such as variability in building size), and urbanization (densely packed domestic structures with streets). Still other sites have yielded evidence for economic specialization (seen in sealings or the existence of a granary). Models can be generated from these data, and then applied to the region to create a working hypothesis of social complexity. Faunal analysis is one tool that is both useful and successful, as this dissertation has shown, at generating such models from an economic perspective that can then be applied to political and social variables.

The case of Helike can now be put to the pig (and faunal) litmus test in order to generate a working model of Early Helladic social complexity. Faunal analysis provides insight into broad economic aspects of the settlement at Helike. The high proportion of pigs in relationship to the other heavily exploited animals, suggests that Helike was a smaller settlement, relatively autonomous in terms of production, with small-scale garden cultivation. Economic specialization may have occurred, but this would probably have been at a household level. It is possible that production was loosely organized at the community level, but the fauna suggest that Helike's interaction with other surrounding settlements was limited.

Looking at the remaining material evidence at Helike further supports the conclusion suggested by the fauna. Architecturally, there is no variability in the size of dwellings. They appear to have been single storied, constructed with stone foundations, mud brick walls, and probably thatched roofs. The absence of roof tiles further suggests a smaller settlement. Limited contact with the wider region is suggested by pottery and by the presence of obsidian. Although there is a high frequency of pigs in the Helike faunal assemblage, there is a noted absence of prime-aged pigs. This may provide a clue as to Helike's economic ties to the region. It is possibly representative of Helike's role as a production site for pigs, and this was the commodity she traded for material goods such as pottery and obsidian.

At Helike, the faunal record and other forms of material evidence compliment one another. They suggest a settlement of relatively autonomous nature, with limited wider, economic contact. Politically and socially, such a site would fit well with Pullen's (2003) chiefdom hypothesis. In a society organized into chiefdoms, there is little centralized political control. However, in chiefdoms, economic specialization and wider economic relationships can exist. Smaller settlements often retain political independence, while still possessing trade contacts. Lerna's probable role as an administrative center, coupled with Tiryns potential role as the same – and perhaps political center, fits well in a chiefdom hypothesis.

Does social complexity exist in Early Helladic Greece? Certainly the beginning of the complexity that marks the Late Bronze Age Mycenaean societies is present. The assignation of Early Bronze Age Greece as a socially complex society depends largely on whether this assignation is parsed out by degrees. Is it as socially complex as Mycenaean Greece? Certainly not. Is it more complex than the Neolithic? Certainly. If social complexity is delineated and identified by the presence of wider economic inter site relationships, central administrative or political centers, some measure of urbanization, and social stratification, then Early Helladic Greece must be socially complex. However, what if all these factors are present, but not at every site? If all components can be found somewhere, then the seeds of social complexity are germinating. This scenario should be marked by wide variability in settlement size, organization, and subsistence – and this is the case. Thus, all evidence suggests that a cohesive political and social network is just beginning in Early Bronze Age Greece. Early Helladic Greece is a socially complex society, but one in its infancy.

> It thus seems as if various types of nucleation appear at times when society is complex enough to encompass several regions. Since nucleation is often a response to economic and demographic factors, the same process will concern a wide area if this is somehow integrated economically. A similar nucleation pattern in EHII/III occurring in several regions of mainland Greece, would then imply an economically integrated area (Schallin 1997:39).

The frequency of pigs at Lerna, Helike, Tsoungiza and Tiryns are an apt illustration of the naissance of social complexity in the Early Bronze Age. Just as there is site variability, there is faunal variability. This faunal variability is an indicator of the nature and function of the settlement. The composition of animals reflects the settlement's economic systems and transitively, its social and political organization. Human animal interactions are a vital component to social organization in every society. As such an integral aspect of human social systems – be they economic or political, they are perhaps the best, most apt indicator of social complexity. Faunal analysis is then the best tool to arrive at economic relationships – producing a well-rounded model on which to base ensuing hypotheses of social complexity in ancient societies. The faunal record, and the frequency of pigs in particular, paints a picture of an Early Helladic Greece with loose inter site economic relationships. There is no clear evidence for wide-spread producer and consumer sites, and thus no clear evidence of centralized political control or centralized social stratification. There may be intra-site social hierarchies, such as at Lerna and Tiryns, but the relatively independent nature of Tsoungiza and Helike, as exemplified by both the fauna and other lines of evidence, mitigates against a politically – and socially- integrated Early Helladic Greece.

This analysis has illustrated the utility of examining non-traditional lines of evidence to approach the traditional study of social complexity. It has shown that faunal analysis, and the analysis of pigs in particular, can be used to address economic aspects of social complexity in ancient settlements. The close and intricate relationship of humans to animals is one that is the hallmark of every human society. This relationship is widely recognized in other areas of the world, but has yet to be so widely accepted as a legitimate framework within which to analyze social complexity in Greece. The successful generation of a model that utilizes pigs as proxy indicators of economic organization to approach larger social questions can be used, with care, on its own, and in conjunction with more traditional archaeological lines of evidence in this region of the world. This model also highlights the utility and necessity of inter-disciplinary study to approach the phenomenon of social complexity, which is as complex and multi-faceted as the variables, and the human agencies, that comprise it.

APPENDIX

Helike – NISP by Species

Specimen	*Pig*	*Cow*	*Sheep/ Goat*	*Deer*	*Horse*	*Human*	*Unidentified*
Antler	-	-	-	1	-	-	-
Horn Core	-	1	4	-	-	-	-
Skull	2	2	1	-	-	-	4
Maxilla	11	-	1	-	-	1	-
Mandible	18	8	10	1	-	-	3
Molar	14	14	15	-	-	2	-
Premolar	10	3	9	-	-	-	-
Canine	7	-	-	-	-	-	-
Incisor	1	1	2	-	-	-	-
Vertebra	7	-	6	-	-	-	3
Rib	7	8	11	-	-	-	-
Pelvis	6	3	4	1	-	-	1
Sacrum	2	3	1	-	-	-	-
Scapula	5	3	9	-	-	-	1
Humerus	9	-	8	-	-	-	-
Radius	6	2	3	1	-	-	-
Ulna	9	3	4	1	-	-	-
Long Bone Shaft	3	6	5	-	-	-	-
Femur	3	2	1	-	-	-	1
Tibia	8	5	9	-	1	-	1
Fibula	1	-	2	-	-	-	-
Metacarpal	1	5	1	-	-	-	-
Astragalus	4	5	1	-	-	-	-
Calcaneus	1	3	3	-	-	-	-
Metapodial	2	9	2	-	-	-	1
Metatarsal	3	3	2	-	-	-	-
Phalange	5	6	2	1	-	-	-
Total NISP	**145**	**95**	**117**	**6**	**1**	**3**	**15**
Assemblage Total	**382**						

Table 1

Observed MNI Frequencies of Bos skeletal specimens from EH Helike

Skeletal Part	*MNI Left*	*MNI Right*	*MNI Unsided*	*MNI (per skeletal portion)*	*MNE*
Cranium	-	-	2	1	1
Horn Core	-	1	-	1	1
Mandible	3	3	1	4	5
M_1	2	2		2	
M^2		2		2	
M_2		1		1	
M^3	1			1	
M_3	2	1		2	
M			3	3	
P			3	3	
I	-	-	1	2	
D Scapula	-	2	1	2	3
M Rib			7	7	8
Pelvis	1	1	1	1	1
Sacrum	-	-	3	3	3
P Radius	1	1		1	2
P Ulna	2(1J)	1		2	3
P Metacarpal		3		3	Mtc 5
D Metacarpal	2			2	
Calcaneus	2	1	-	2	3
P Femur	-	1	1	1	2
P Tibia	1	1		1	Tibia 3
D Tibia	1	2		2	
Astragalus	2	2	1	2	5
Metapodial	1	1	7	7	9
P Metatarsal	1		1	1	Mt 1
D Metatarsal			1	1	
P 1st Phalanx		1		1	1
D 2nd Phalanx			2	2	2nd Phalanx 2
P 2nd Phalanx	1			1	
Phalanx	-	-	2	1	2
Total				**MNI 7**	**MNE 60**

Table 2

Observed MNI Frequencies of Ovicaprid skeletal specimens from EH Helike

Skeletal Part	*MNI Left*	*MNI Right*	*MNI Unsided*	*MNI(per skeletal portion)*	*MNE*
Skull			1	1	1
Horn Core	1		3	3	4
Maxilla		1		1	1
Mandible	2	6	2	6	10
M^1		2			
M_1	1	1	1		
M_2	3	3			
M^3		1			
M_3		1			
M			2		
P^4		1			
P_4		1			
P_2		2			
P^2		1			
P_3		2	1		
P^3		1			
I	1		1		
Vertebra			6	1	6
D Scapula	2	3	4	4	9
M Rib			11	1	11
Pelvis	2	1	1	2	4
Sacrum			1	1	1
D Humerus	4	2	2	4	8
M Radius	1		1	2	Radius 3
P Radius	1			1	
P Ulna	2			2	Ulna 4
M Ulna			2	1	
P Fibula			2	1	2
P Metacarpal	1			1	1
P Femur	1			1	1
P Tibia	2		2	2	Tibia 6
M Tibia	1			1	
D Tibia	3	1		3	
Astragalus		1		1	1
Calcaneus	1	2		2	3
Metapodial			2	1	2
P Metatarsal		1	1	1	2
Long bone shaft	-	-	5	1	5
Total				**MNI 6**	**MNE 87**

Table 3

Observed MNI Frequencies of sus skeletal specimens from EH Helike

Skeletal Part	*MNI Left*	*MNI Right*	*MNI Unsided*	*MNI(per skeletal portion)*	*MNE*
Cranium	-	-	2	1	2
Maxilla	9	-	2	9	11
Mandible	5	7	6	7	18
M^1	3	-	-	3	
M_1	1	-	-	1	
M^2	3	-	-	3	
dM^2	1	-	-	1	
M^3	2	-	1	2	
M_3	1	1	-	1	
M	-	-	1	1	
P^4	1	-	1	1	
P_4	1	-	-	1	
P_3	2	-	-	2	
dP_1	1	-	-	1	
dP_2	1	-	-	1	
dP^2	1	-	-	1	
dP_3	1	-	-	1	
dP^3	1	-	-	1	
C^1	2(M,F)	-	1 upper/lower?	2	
C_1	1(J)	2(M,F)	1	3	
I	-	-	1(lower)	1	
Atlas	-	-	2	2	2
Vertebra	-	-	5	1	5
M Scapula	-	-	1	1	
D Scapula	4	-	-	4	4 Scapula
M Rib	-	-	7	1	7
Sacrum	-	-	2	2	2
Pelvis	1	1	4	2	6
M Humerus	-	-	2	2	
P Radius	2	1(J)	1	3	
P Ulna	1(J)	2(1J)	3(J)	2	7 ulna
M Ulna	-	-	2	2	
P Metacarpal	1	-	-	1	1
D Femur	2(1J)	-	1(J)	2	3
P Tibia	1(J)	1(nearly whole)(J)	2(J)	2	4 tibia
D Tibia	-	-	2(J)	2	
M Tibia	-	-	2(1J)	2	
P Fibula	-	-	1	1	1
Long bone shaft	-	-	3	1	3
Calcaneus	-	-	1	1	1

Table 4 **(Continued Overleaf…)**

Skeletal Part	MNI Left	MNI Right	MNI Unsided	MNI(per skeletal portion)	MNE
M Tibia	-	-	2(1J)	2	
P Fibula	-	-	1	1	1
Long bone shaft	-	-	3	1	3
Calcaneus	-	-	1	1	1
Astragalus	1	1	2	2	4
Metapodial	-	-	2	2	2
P Metatarsal	1	-	-	1	Mt 2
D Metatarsal	2(1J)	-	-	2	
1st Phalanx	1	2	-	2	3
2nd Phalanx	1	-	-	1	1
Phalange	-	-	1(J)	1	1
Total				**MNI 9**	**MNE99**

Table 4

Assemblage		*Pig*	*Cow*	*Sheep/Goat*	*Other mammals*	*Total*
FN-EH Tsoungiza	MinAU %	359 37.7%	133 14.0%	415 43.6%	45 4.7%	**952**
EH Tiryns	NISP %	1157 26.0%	1215 27.3%	1953 44.0%	118 2.7%	**4443**
EH(Lerna III/IV)	NISP %	1432 30.2%	1322 27.9%	1637 34.5%	347 7.3%	**4738**
EH Helike	NISP %	145 37.9%	95 24.9%	117 30.6%	25 6.5%	**382**

Table 5: **Taxonomic Composition of EH assemblages from Tsoungiza (Halstead in press), Tiryns (von den Driesch and Boessneck 1990), Lerna (Gejvall 1969:10) and Helike (Fillios, unpublished)**

Skeletal Part	*Pig*	*Cow*	*Sheep/Goat*
Cranial	31 / 31.3%	7 / 12%	16/ 18%
Axial	22 / 22%	12 / 20%	22 / 25%
Appendicular	31 / 31.3%	13 / 21	38 / 44%
Foot	15 / 15%	28 / 47%	11 / 13%
Total	**99**	**60**	**87**

Table 6. **Skeletal Part Comparison – Helike (MNE / %MNE) (Fillios, unpublished)**

Skeletal Part	*Pig*	*Cow*	*Sheep/Goat*
Cranial	52 / 20.5%	16 / 16.6%	56 / 18.7%
Axial	22 / 8.7%	7 / 7.3%	24 / 8%
Appendicular	136 / 53.7%	32 / 33.3%	137 / 45.8%
Foot	43 / 17%	41 / 42.7%	82 / 27.4%
Total	**253**	**96**	**299**

Table 7. **Skeletal Part Comparison – Tsoungiza (MinAU / %MinAU) (after Halstead, in press)**

Skeletal Part	*Pig*	*Cow*	*Sheep/Goat*
Cranial	714 / 50%	438 / 33%	828 / 51%
Axial	44 / 3%	111 / 8%	50 / 3%
Appendicular	556 / 39%	299 / 23%	557 / 34%
Foot	118 / 8%	474 / 36%	202 / 12%
Total	**1432**	**1322**	**1637**

Table 8. **Skeletal Part Comparison- Lerna (NISP / % NISP) (after Gejvall 1969)**

Skeletal Part	*Pig*	*Cow*	*Sheep/Goat*
Cranial	370 / 32%	242 / 19.9%	381 / 19.5%
Axial	291 / 25.2%	370 / 30.5%	381 / 19.5%
Appendicular	402 / 34.7%	424 / 34.9%	926 / 47.4%
Foot	94 / 8.1%	179 / 14.7%	265 / 13.6%
Total	**1157**	**1215**	**1953**

Table 9. **Skeletal Part Comparison- Tiryns (NISP / % NISP) (After Table 7a von den Driesch and Boessneck 1990:95)**

Species	*Under 6 mo.*	*6-12 mo.*	*12-24 mo.*	*Over 24 mo.*	**Total (N)**
Pig	-	4 / 20%	4 / 20%	12 / 60%	**20 / 100%**
Sheep/Goat	-	3 / 15%	5 / 25%	12 / 60%	**20 / 100%**
Cattle	-	-	2 / 15%	11 / 85%	**13 / 100%**
Total (N)	**-**	**7**	**11**	**35**	**53**

Table 10. **EH Helike – Dental Ageing Data NISP/%NISP (Fillios, unpublished)**

Species	Under 6 mo.	6-12 mo.	12-24 mo.	Over 24 mo.	Total (N)
Pig	-	45 / 16%	56 / 20%	178 / 64%	279 / 100%
Sheep/Goat	-	57 / 15%	66 / 18%	247 / 67%	370 / 100%
Cattle	-	4 / 5%	11 / 14%	65/ 81%	80 / 100%
Total (N)	-	106	133	490	729

Table 11. **EH II/III Lerna – Dental Ageing Data NISP/%NISP (after Gejvall 1969)**

Species	Under 6 mo.	6-12 mo.	12-24 mo.	Over 24 mo.	Total (N)
Pig	-	9 / 41%	8 / 36%	5 / 23%	22 / 100%
Sheep/Goat	-	7 / 30%	3 / 13%	13 / 57%	23 / 100%
Cattle	-	1 / 20%	2 / 40%	2 / 40%	5 / 100%
Total (N)	-	17	13	20	50

Table 12. **EH Tsoungiza – Dental Ageing Data MinAU/ %MinAU (after Halstead, in press)**

Species	Under 6 mo.	6-12 mo.	12-24 mo.	Over 24 mo.	Total (N)
Pig	-	5 / 28%	6 / 33%	7 / 39%	18 / 100%
Sheep/Goat	5 / 28%	2/ 11%	11 / 61%	-	18 / 100%
Cattle	-	-	2 / 15%	11 / 85%	13 / 100%
Total (N)	5	7	19	18	49

Table 13. **EH Tiryns – Dental Ageing Data NISP/%NISP (after von den Driesch and Boessneck 1990)**

Element	*Pig*	*Cow*	*Sheep/Goat*	*Total (N)*
Mandible	3	1	-	**4**
Scapula	-	-	1	**1**
D Humerus	3	-	-	**3**
M Humerus	-	-	1	**1**
P Radius	1	-	-	**1**
P Ulna	2	1	-	**3**
D Femur	1	-	-	**1**
D Tibia	-	-	1	**1**
Pelvis	-	2	-	**2**
Rib	-	2	1	**3**
Calcaneus	-	1	1	**2**
Astragalus	-	-	1	**1**
Metacarpal	-	1	-	**1**
Metapodial	-	1	-	**1**
Total	10	9	6	25

Table 14. **EH Helike – Cut mark frequency/skeletal element**

Assemblage		*Pig*	*Cow*	*Sheep/Goat*	*Other Mammals*	*Total*
LH Lerna	NISP %	500 37.8%	169 12.8%	508 38.4%	146 11.0%	**1323**
LH Pylos	NISP %	1944 32.5%	1196 20.0%	2762 46.1%	87 1.5%	**5989**
LH Nichoria	NISP %	759 34.5%	329 14.9%	1079 49.0%	34 1.5%	**2201**
LH Midea	NISP %	121 19.7%	136 22.2%	343 56%	13 2.1%	**613**
LH Tsoungiza (feasting deposit)	MinAU %	16 16%	54 54%	30 30%	/	**100**
LH Tiryns	NISP %	8636 21.9%	14599 37.1%	16138 41.0%	/	**39373**
LH Eleusis	NISP %	40 18.4%	32 14.7%	141 65.0%	4 1.8%	**217**
LH Agios Konstantinos (Mycenaean sanctuary)	NISP **%**	50 38.8%	2 1.6%	75 58.1%	2 1.6%	**129**

Table 15. **Taxonomic Composition of LH Faunal Assemblages** (Lerna Gejvall 1969, Pylos Nobis 1993, Nichoria Sloan and Duncan 1978, Midea Reese 1998, Tsoungiza Dabney et al. 2004, Tiryns von den Driesch & Boessneck 1990, Eleusis Cosmopoulos et al. 2003, A.Konstantinos Hamilakis & Konsolaki 2004)

Site	*% Sheep*	*%Goat*	*% Cow*	*%Pig*	*Total identified*	*Reference*
Achilleion	-	86	4	9	2256	**Bökönyi 1989**
Argissa	84	0	5	11	2178	**Boessneck 1962**
Agios Petros	-	85	7	8	-	**Schwartz 1982**
Knossos	58	8	16	17	1999	**Jarman & Jarman 1968**
Lerna	-	64	12	24	141	**Gejvall 1969**
Nea Nikomedeia	-	71	15	15	439	**Higgs 1962**
Otzaki	-	52	31	17	297	**Boessneck 1955**
Prodhromos 1-2	48	9	29	14	1362	**Halstead & Jones 1980**
Prodhromos 3	40	10	38	12	285	**Halstead & Jones 1980**
Servia	-	66	17	17	-	**Watson 1979**
Sesklo	-	67	13	20	447	**Schwartz 1982**

Table 16. **Taxonomic Composition of EN Faunal Assemblages (after Halstead 1992)**

EH Helike – Pig Metrical Data (Teeth)

SPECIMEN	TYPE	MEASUREMENT (mm) 'B'	MEASUREMENT (mm) 'L'
CANINE	Lower	19.74	
		5.51	11.51
		19.80	
	Upper	8.12	12.60
INCISOR	Lower	7.47	6.09
DECIDUOUS	dP_1	3.18	7.97
	dP_2	4.08	10.75
	dP_3	8.69	18.77
	dP^2	8.38	12.84
	dP^3	11.60	13.41
PRE-MOLAR	P_3	5.09	10.3
			11.57
	P_4	8.40	13.46
		12.28	13.35
	P^4	11.94	11.04
MOLAR	M_1	10.64	14.24
	M^1	13.93	17.70
		10.84	14.30
		12.47	14.11
		13.48	17.95
	M^2	17.30	20.00
		12.51	16.88
		16.55	21.39
		12.93	17.08
		17.05	18.45
	M^3	18.73	28.56
		18.09	31.32
	M_3	15.50	30.45
		15.82	
MANDIBLE		GOC = 77.37	Length of Molar Row = 62.57

Table 17

EH Helike – Pig Metrical Data (Post Cranial)

SPECIMEN	*TYPE*	*MEASUREMENT (millimeters)*									
ATLAS		**H** = 44.06 (preserved, some missing)									
		H = 46.81 (preserved, some missing)									
		MEASUREMENT (millimeters)									
		SLC	**GLP**	**LG**	**BG**	**DPA**	**SDO**	**Bd**	**Bp**	**BT**	**SD**
SCAPULA		23.19									
		22.8									
		24.2	34.91	26.06	22.82						
		21.96	31.7	23.57	19.07						
RADIUS									24.02		
									27.97		
									26.78		
ULNA						31.27					
						31.78					
						35.2	26.65				
HUMERUS								38.18		30.88	
TIBIA									44.0		18.39
	TYPE	*MEASUREMENT (millimeters)*									
			GL	**Gli**	**GLm**	**Glpe**	**B**	**Bd**	**Bp**		
METACARPAL	(III)								17.27		
METATARSAL	(III)							15.55			
	(III)		80.22 preserved				13.9		15.73		
	(III)								16.37		
ASTRAGALUS				38.54	33.78						
				29.28	28.81 broken						
PHALANGE								12.26			
	(1st)		32.53 preserved						15.9		
	(1st)								15.63		
	(2nd)		22.94			21.13		14.0	15.19		
	(1st)		39.62					15.33	16.87		

Table 18

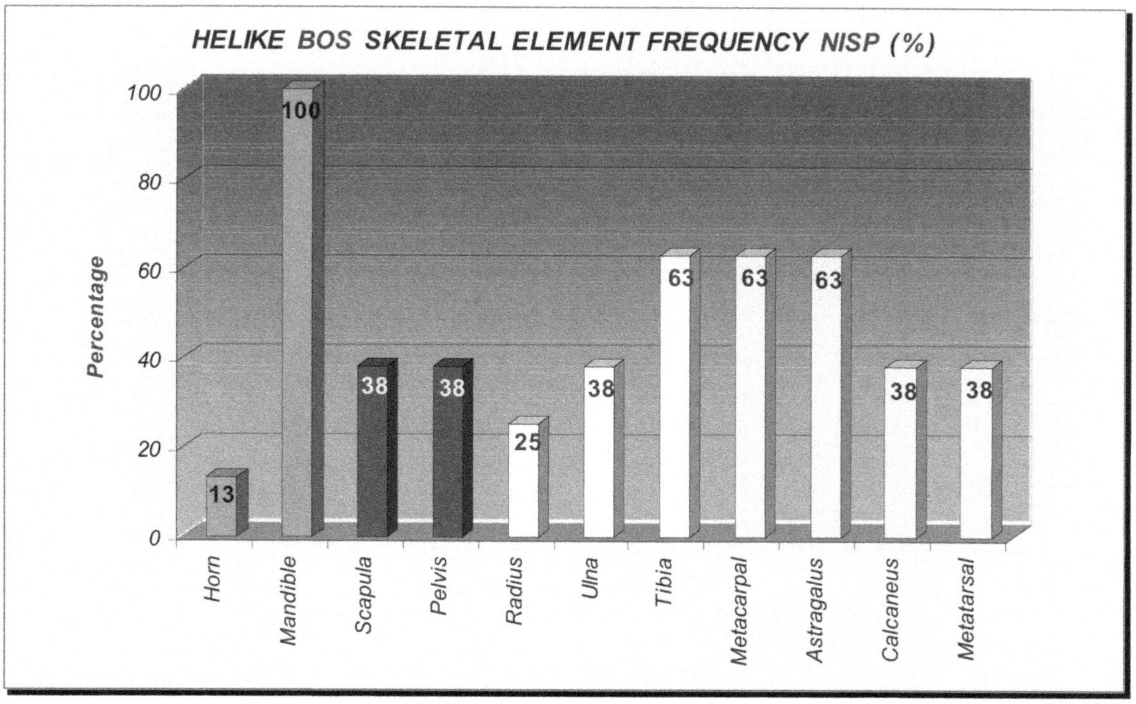

Figure 1

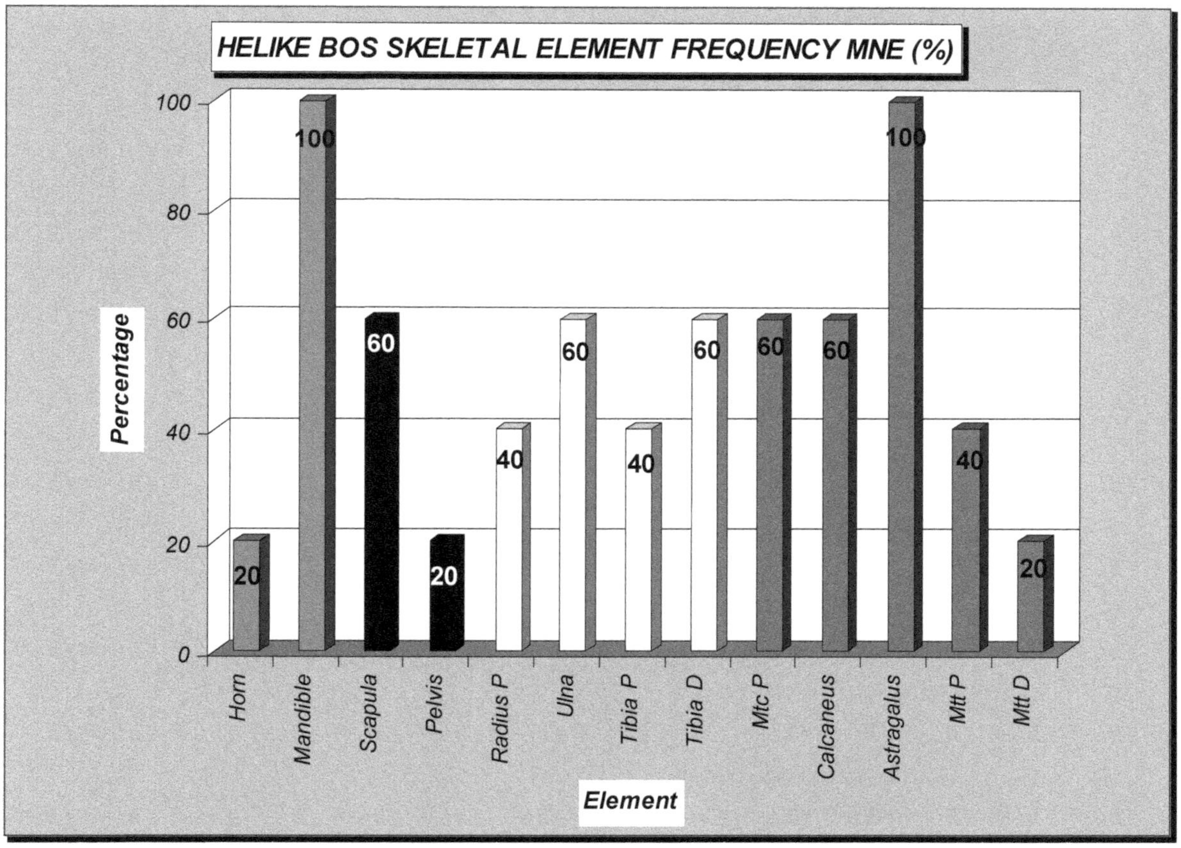

Figure 2

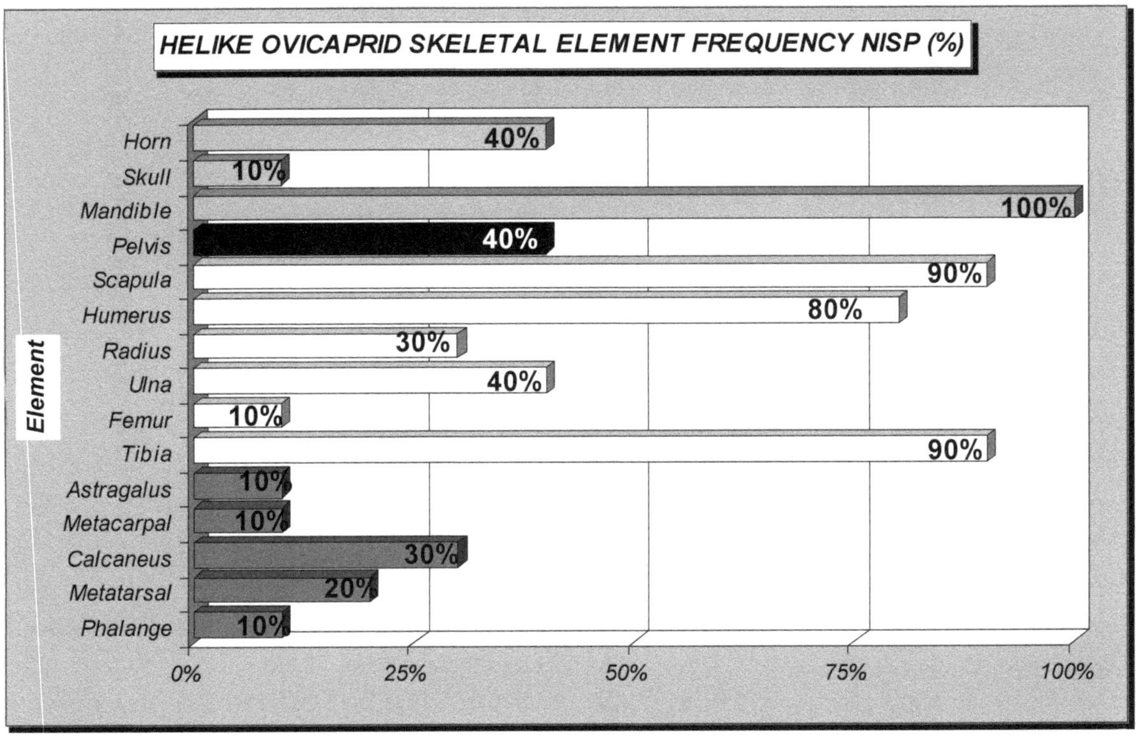

Figure 3

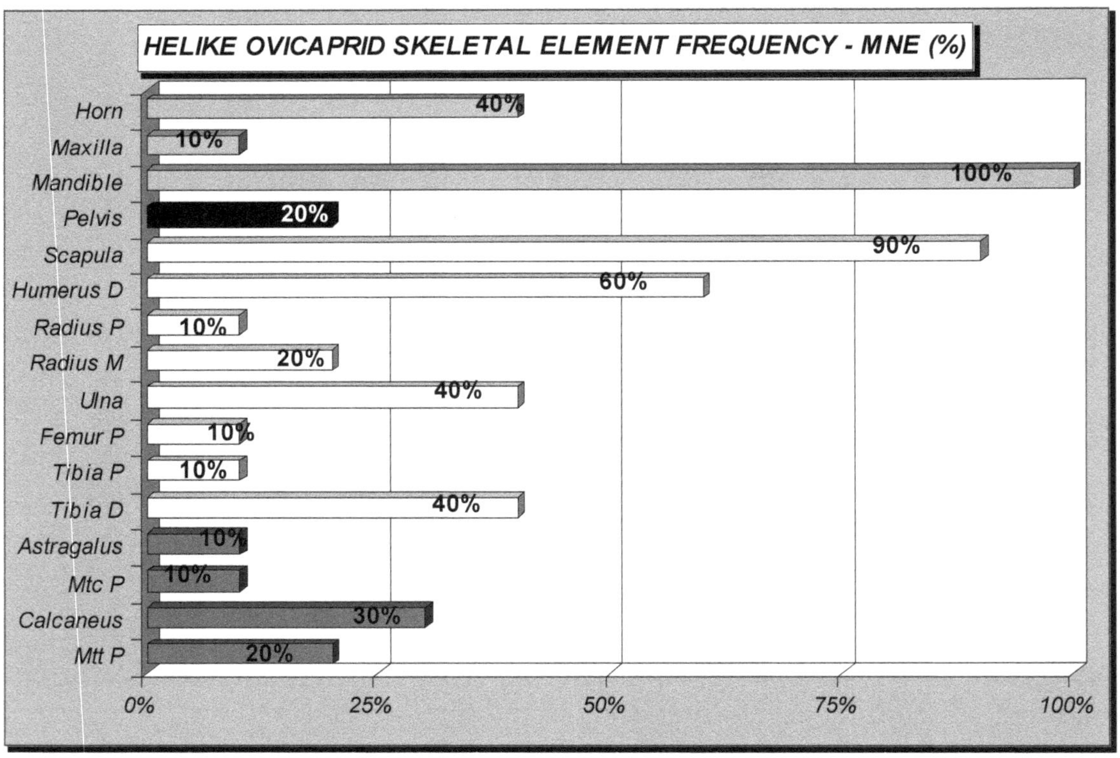

Figure 4

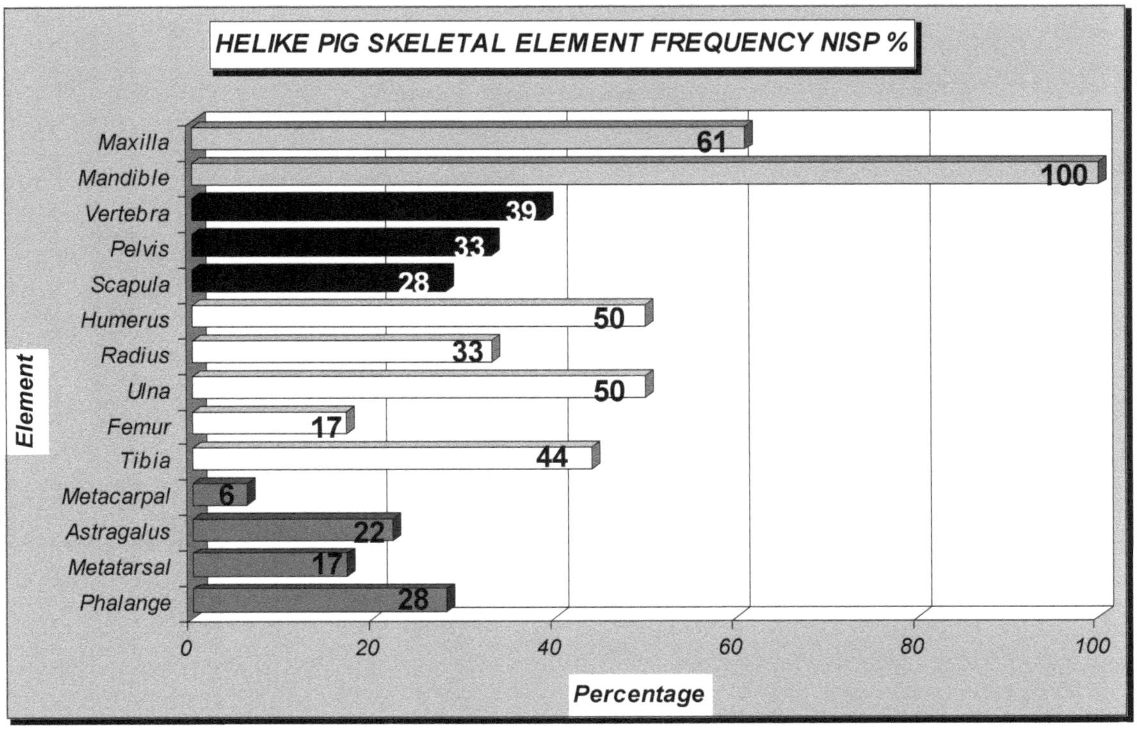

Figure 5

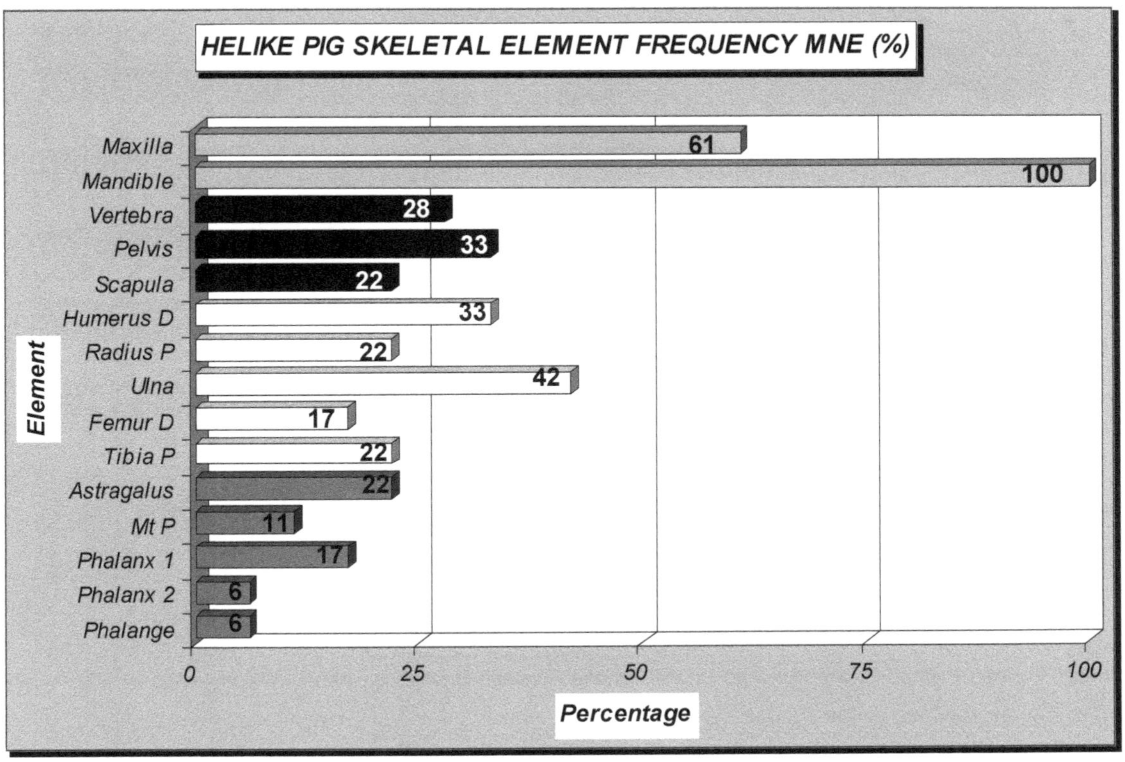

Figure 6

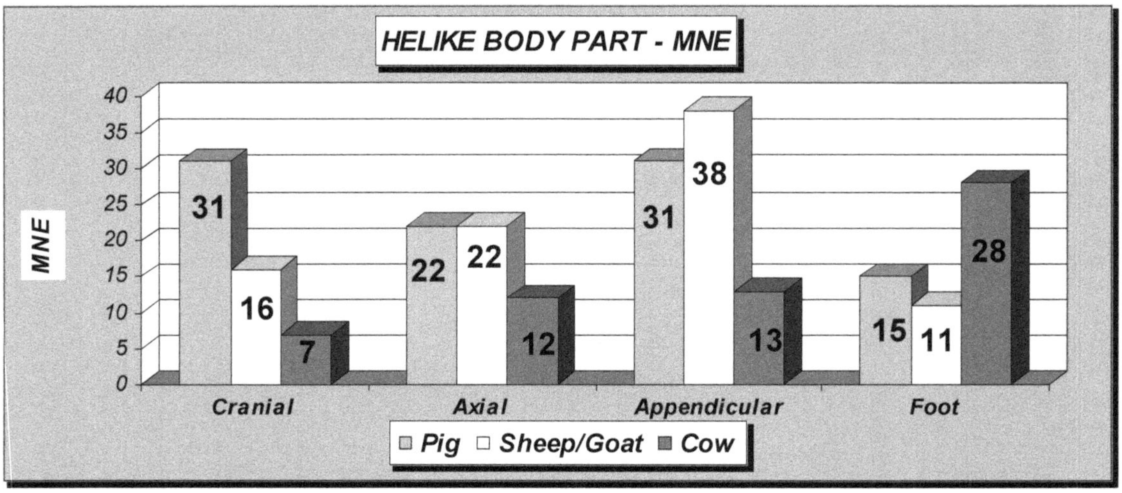

Figure 7. **Helike Body part representation (MNE)**

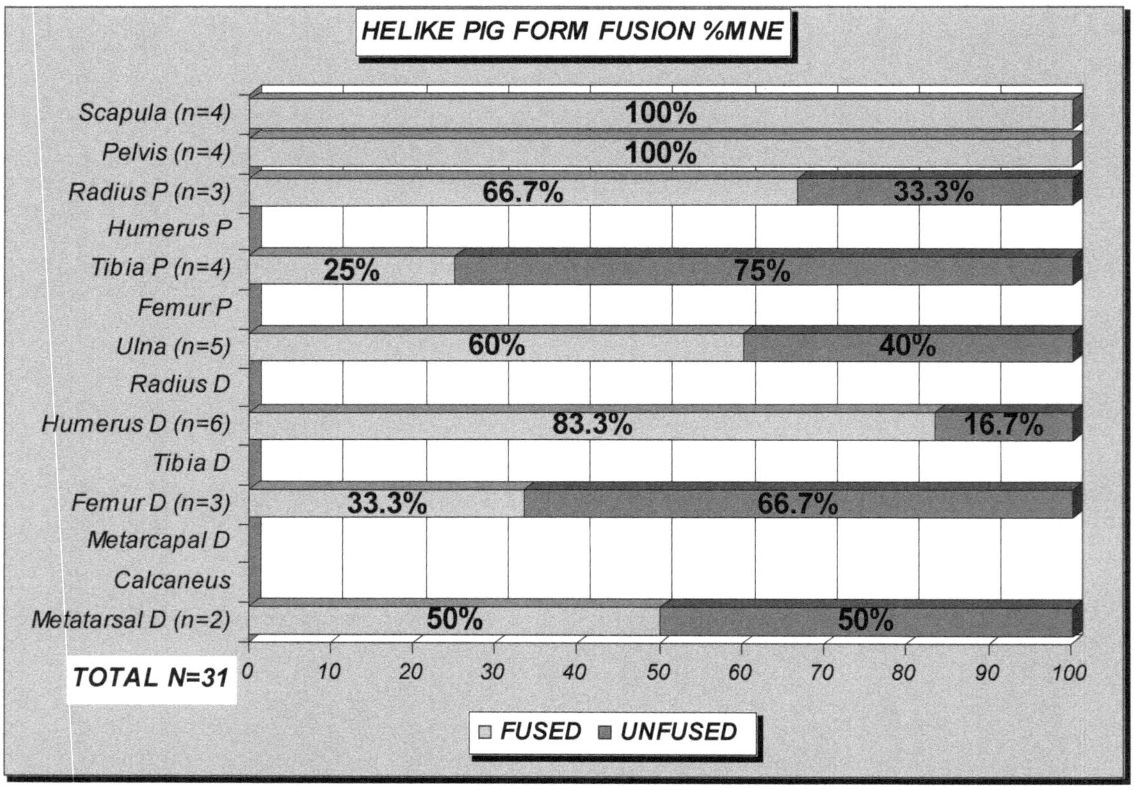

Figure 8. **Helike Pig Form Fusion Data (% MNE)**

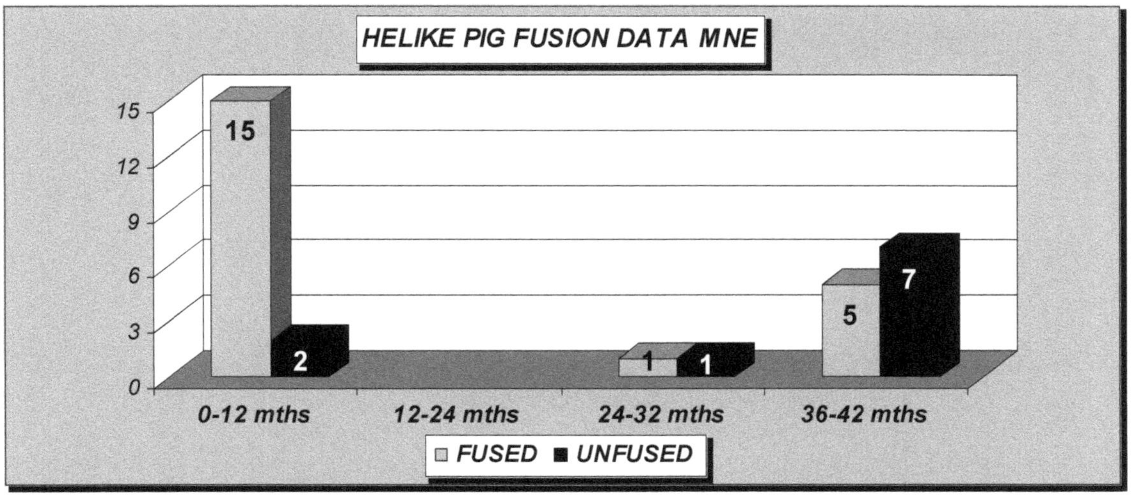

Figure 9. **Helike Pig Fusion Data (Month of Fusion MNE)**

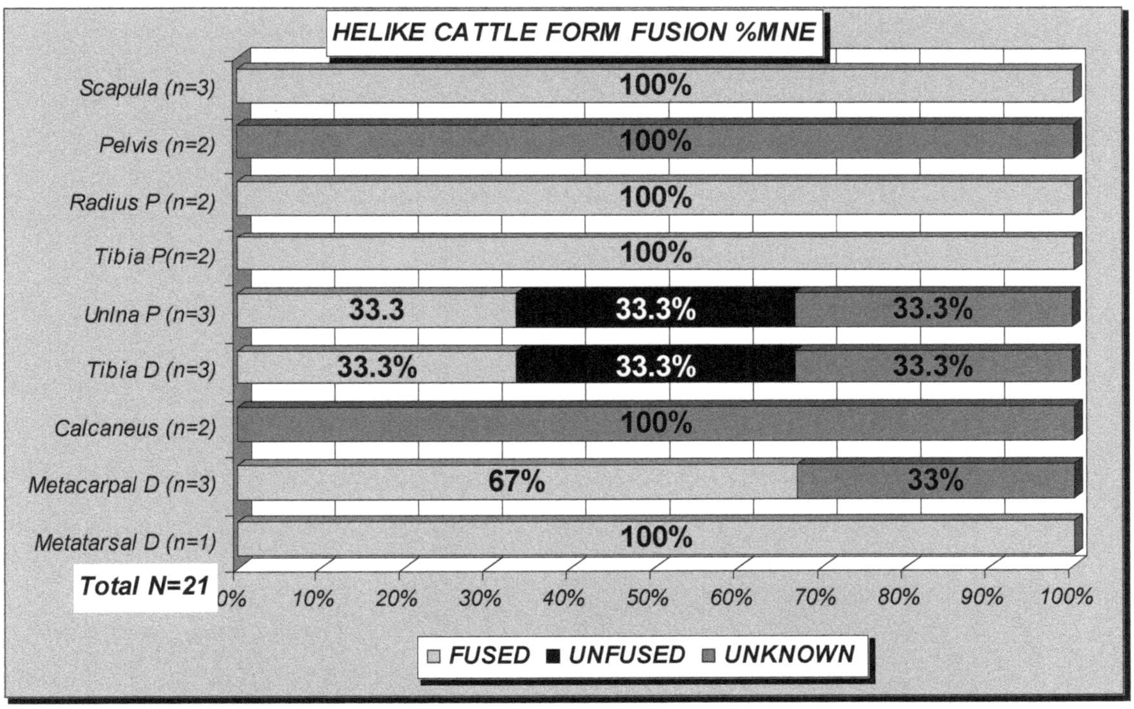

Figure 10. **Helike Cattle Form Fusion (%MNE)**

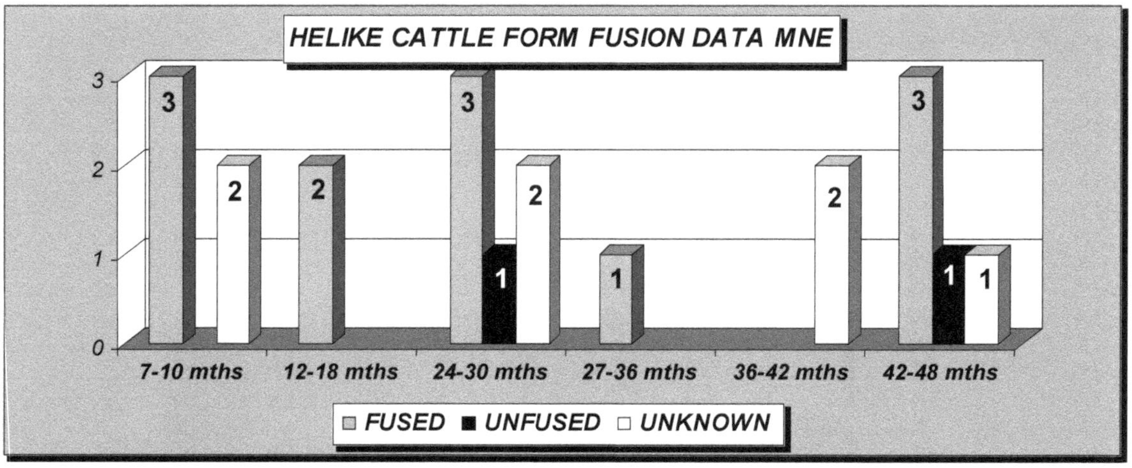

Figure 11. **Helike Cattle Month of Fusion Data (MNE)**

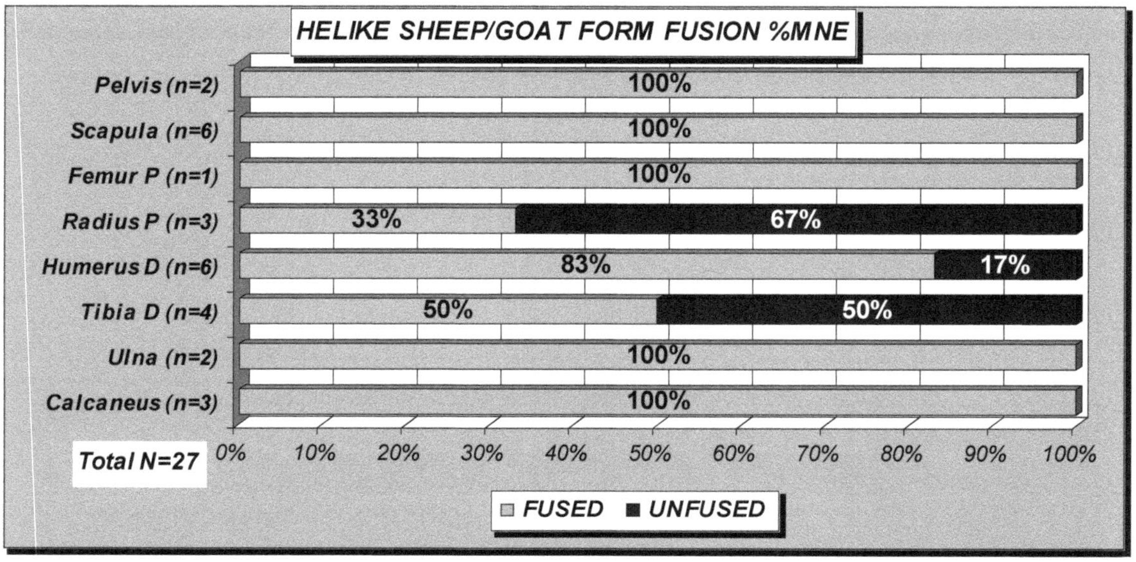

Figure 12. **Helike Sheep/Goat Form Fusion (%MNE)**

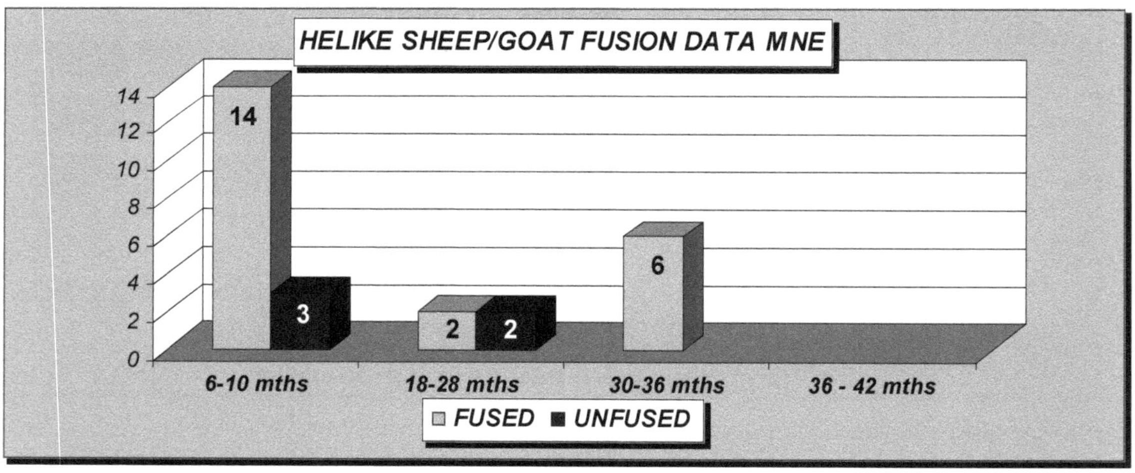

Figure 13. **Helike Sheep/Goat Month of Fusion (MNE)**

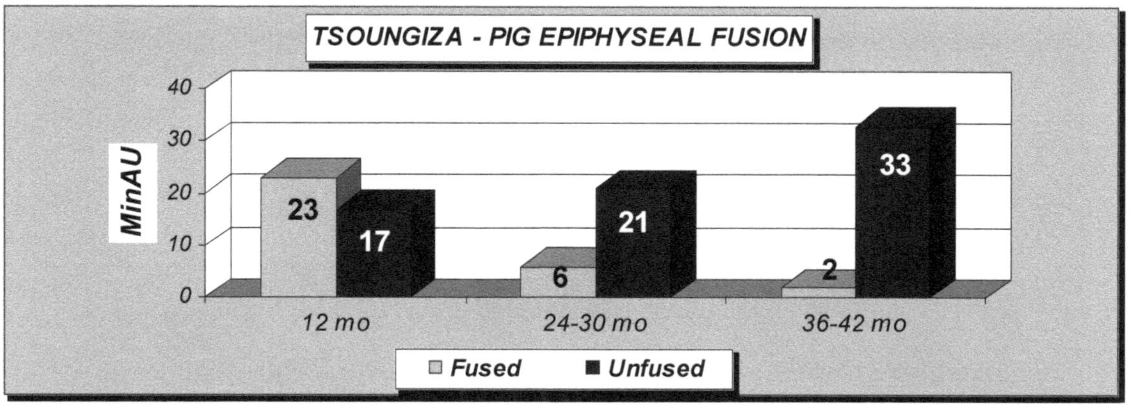

Figure 14

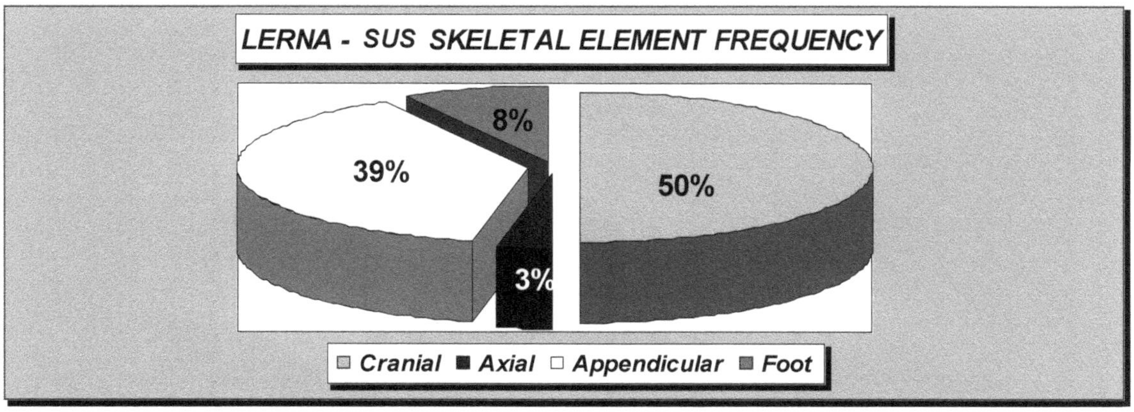

Figure 15

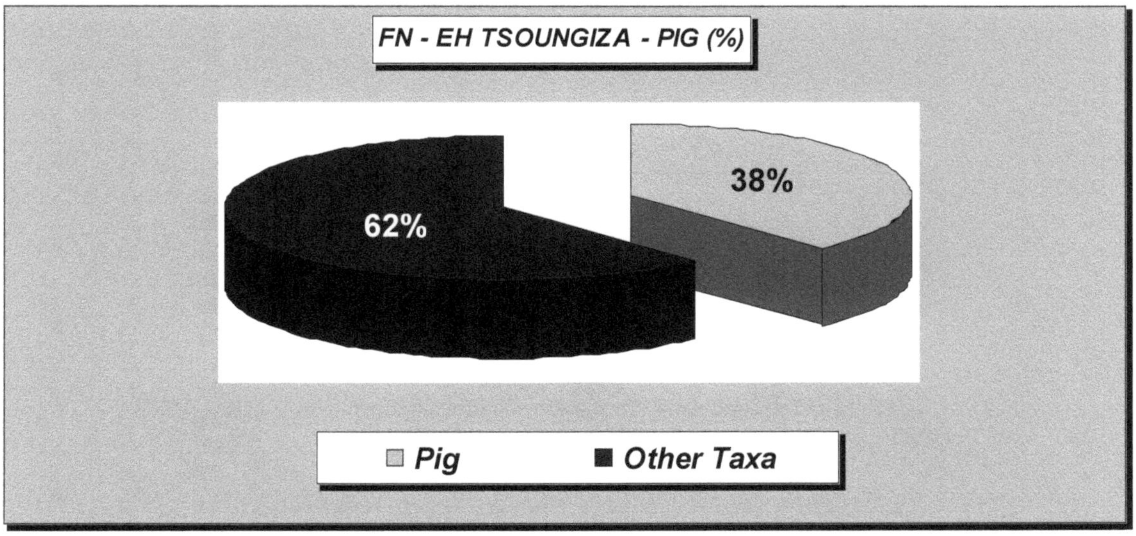

Figure 16

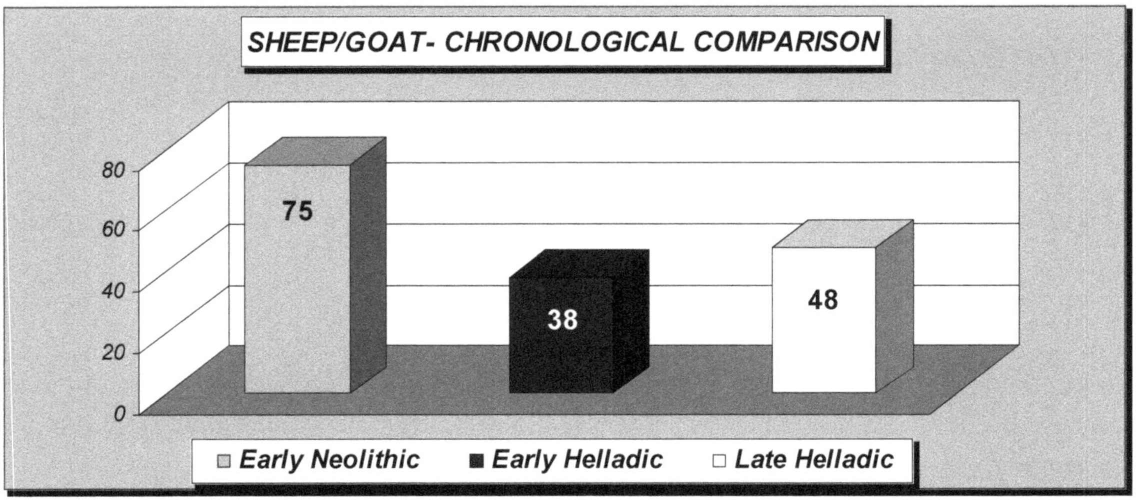

Figure 17. **Chronological comparison of the major domesticates in Greece - Sheep/Goat**

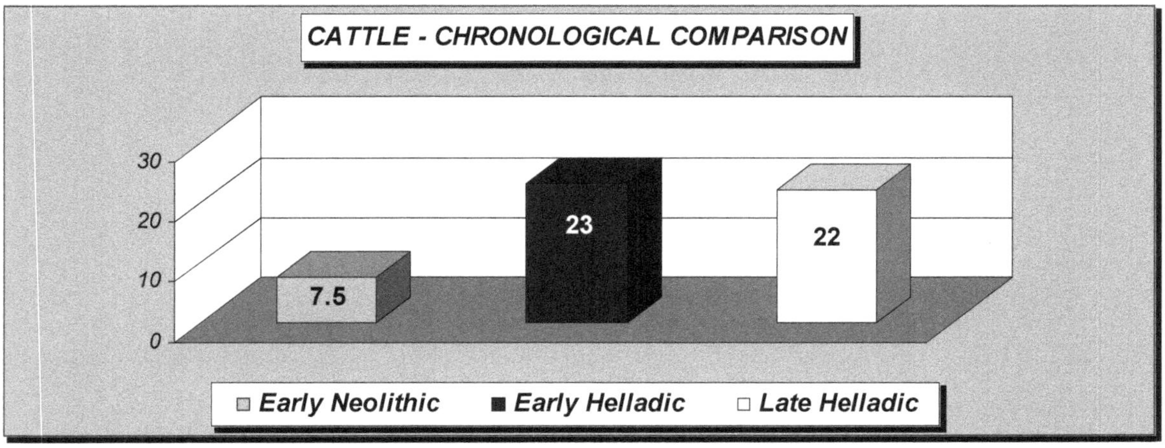

Figure 18. **Chronological comparison of the major domesticates in Greece – Cattle**

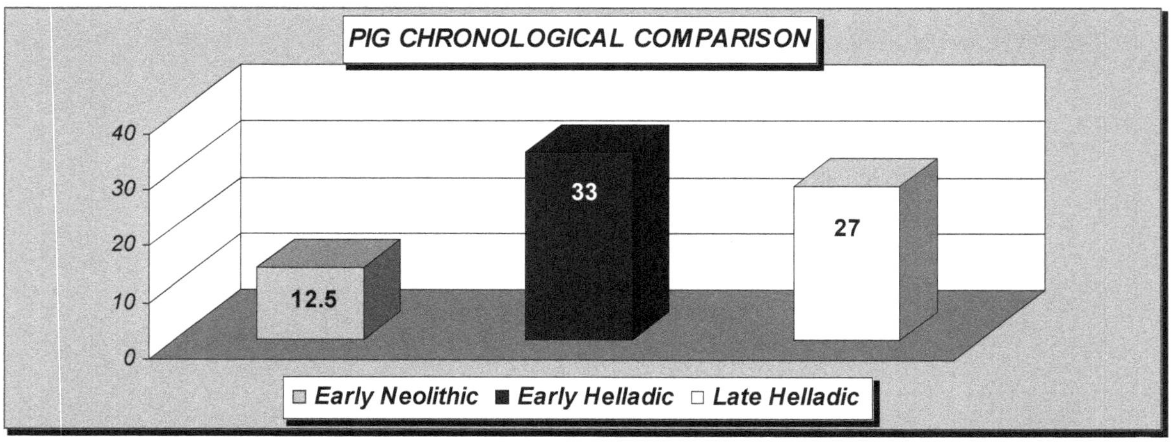

Figure 19. **Chronological comparison of the major domesticates in Greece – Pig**

BIBLIOGRAPHY

Abe, Y., C. W. Marean, P. J. Nilssen, Z. Assefa, and E. C. Stone. 2002. The Analysis of Cutmarks On Archaeofauna: A Review And Critique Of Quantification Procedures, And A New Image Analysis GIS Approach. *American Antiquity* 67:643-663.

Albarella, U., and D. Serjeantson. 2002. "A Passion for Pork: Meat Consumption at the British Late Neolithic Site of Durrington Walls," in *Consuming Passions and Patterns of Consumption*. Edited by P. Miracle and N. Milner, pp. 33-49. Oxford: McDonald Institute for Archaeological Research.

Alcock, S. E., and J. F. Cherry. Editors. 2004. *Side-by-Side Survey: Comparative Regional Studies in the Mediterranean World*. Oxford: Oxbow Books.

Amato, C. 2000. *The Wild Boar: History, Husbandry, The Hunt: A Complete Study of Sus scofa*.

Andersen, H. D., W. Horsnaes Helle, S. Houby-Nelson, and A. Rathje. Editors. 1997. *Urbanization in the Mediterranean in the 9th to 6th Centuries B.C. Acta Hyperborea 7 Danish Studies in Classical Archaeology*. Copenhagen: Museum Tusculanum Press, University of Copenhagen.

Asher, N. S., and D. Small. Editors. 1997. *The Archaeology of Israel: Constructing the Past, Interpreting the Present. Journal for the Study of the Old Testament Supplement Series 237*. Sheffield: Sheffield Academic Press LTD.

Attas, M., J. M. Fossey, and L. Yaffe. 1987. An Archaeometric Study of Early Bronze Age Pottery Production in Argolis and Korinthia (Corinthia), Greece. *Journal of Field Archaeology* 14:77-90.

Aufrecht, W. E., N. A. Mirau, and S. W. Gauley. Editors. 1997. *Urbanism in Antiquity: From Mesopotamia to Crete. Journal for the Study of the Old Testament Supplement Series 244*. Sheffield: Sheffield Academic Press.

Bailey, G. N. Editor. 1997. *Klithi: Paleolithic Settlement and Quaternary Landscapes in Northwest Greece. Vol. 1: Excavation and intra-site analysis at Klithi*. McDonald Institute Monographs. Cambridge: McDonald Institute for Archaeological Research.

Bailey, G. N., E. Adam, E. Panagopoulou, C. Perles, and K. Zachos. Editors. 1999. *The Paleolithic Archaeology of Greece and Adjacent Areas*. British School at Athens Studies 3. London: The Managing Committee, British School at Athens.

Bailey, G. N., Carter, P., Gamble, C., Higgs, H. 1983. "Epirus Revisited: Seasonality and Inter-Site Variation in the Upper Paleolithic of North-West Greece," in *Hunter-Gatherer Economy in Prehistory*. Edited by G. N. Bailey, pp. 64-78. Cambridge: Cambridge University Press.

Baker, J. R. 1984. "The Study of Animal Diseases with Regard to Agricultural Practices and Man's Attitude to His Animals," in *Animals and Archaeology: 4. Husbandry in Europe*, vol. 227, BAR International Series. Edited by C. Grigson and J. Clutton-Brock, pp. 253-257. Oxford: BAR.

Bar-Oz, G., and T. Dayan. 2003. Testing the use of multivariate inter-site taphonomic comparisons: the faunal analysis of Hefzibah in its Epipaleolithic cultural context. *Journal of Archaeological Science* 30:885-900.

Barringer, J. 2001. *The Hunt in Ancient Greece*. Baltimore: The Johns Hopkins University Press.

Becker, C. 1986. Kastanas, Ausgrabungen in einem Siedlungshugel der Bronze und Eisenzeit Makedoniens 1975-1979. Die Tierknochenfunde. *Prahistorische Archaologie in Sudosteuropa* 5.

Beech, M. 2000. "Preliminary Report on the Faunal Remains From an 'Ubaid Settlement on Dalma Island, United Arab Emirates." in *Archaeozoology of the Near East IV,* Fourth International Symposium on the Archaeozoology of Southwestern Asia and Adjacent Areas, Groningen, The Netherlands, 2000, pp. 68-78.

Behrensmeyer, A. K. 1975. Taphonomy and Paleoecology in the Hominid Fossil Record. *Yearbook of Physical Anthropology* 19:36-50.

Beirs, W. R. 1996. *The Archaeology of Greece*. Ithaca, NY: Cornell University Press.

Bendall, L. 2004. "Fit For a King? Hierarchy, Exclusion, Aspiration and Desire inthe Social Structure of Mycenaean Banqueting," in *Food, Cuisine and Society in Prehistoric Greece*. Edited by P. Halstead and J. C. Barrett, pp. 105-135. Oxford: Oxbow Books.

Berry, R. J. 1969. "The Genetical Implications of Domestication in Animals," in *The Domestication and Exploitation of Plants and Animals*. Edited by P. J. Ucko and G. W. Dimbleby, pp. 207-217. Chicago: Aldine Publishing Company.

Biddick, K. 1984. "Pig Husbandry on the Peterborough Abbey Estate From the Twelfth to the Fourteenth Century A.D.," in *Animals and Archaeology: 4. Husbandry in Europe*, vol. 227, BAR International Series. Edited by C. Grigson and J. Clutton-Brock, pp. 161-177. Oxford: B.A.R.

Binford, L. R. 1981. *Bones: Ancient Men and Modern Myths*. Studies in Archaeology. New York: Academic Press.

—. 1984. *Faunal Remains from Klasies River Mouth*. Orlando: Academic Press.

Bintliff, J., and A. M. Snodgrass. 1985. The Cambridge/Bradford Boeotian Expedition: the First Four Years. *Journal of Field Archaeology* 12:123-161.

Blumenshine, R. J., C. W. Marean, and S. D. Capaldo. 1996. Blind Tests of Inter-analyst Correspondence and Accuracy in the Identification of Cut Marks, Percussion Marks, and Carnivore Tooth Marks on Bone Surfaces. *Journal of Archaeological Science* 23:493-507.

Boessneck, J. 1956. Zu den Tierknochenfunden aus neolithischen Siedlungen Thessaliens. *Bericht der Romisch-Germanischen Kommission* 36:1-51.

—. 1962. "Die Tierreste aus der Argissa-Magula vom prakeramischen Neolithikum bis zur mittlere Bronzezeit.," in *Die deutchen Ausgrabungen auf der Argissa-Magula in Thessalien, I. Das prakeramische Neolithickum sowie die Tier- und Pflanzenreste, Beitrage zur ur- und fruhgeschichtlichen Archaologie des Mittelmeer-Kulturraumes 2*. Edited by V. Milojcic, J. Boessneck, and M. Hopf, pp. 27-99. Bonn: Rudolf Habelt.

—. 1965. "Die jungleistozanen Tierknochenfunde aus dem Peneiostal bei Larissa in Thessalien," in *Palaolithikum um Larissa in Thessalien, Beitrage zu ur- und fruhgeschichtlichen Archaologie des Mittelmeer-Kulturraumes 1*. Edited by V. Milojcic, J. Boessneck, D. Jung, and H. Schneider, pp. 42-60. Bonn: Habelt.

Boessneck, J., and A. von den Driesch. 1981. Reste exotischer Tiere aus dem Heraion auf Samos. *Mitteilungen des Deutschen Archeologischen Instituts, Athenische Abteilung* 96:245-248.

Bokonyi, S. 1969. "Archaeological Problems and Methods of Recognizing Animal Domestication," in *The Domestication and Exploitation of Plants and Animals*. Edited by P. J. Ucko and G. W. Dimbleby, pp. 219-229. Chicago: Aldine Publishing Company.

—. 1974. *History of Domestic Mammals in Central and Eastern Europe*. Budapest: Akademiai Kiado.

—. 1986. "Faunal Remains," in *Excavations at Sitagroi: A Prehistoric Village in Northeast Greece*, vol. 1, Monumenta Archaeologica volume 13. Edited by C. Renfrew, M. Gimbutas, and E. S. Elster, pp. 63-132. Los Angeles: Regents of the University of California.

—. 1989. "Animal Remains," in *Achilleion, A Neolithic Settlement in Thessaly, Greece, 6400-5600 B.C.* Edited by M. Gimbutas, S. Winn, and D. Shimabuku, pp. 315-332. Los Angeles: Institute of Archaeology, Monumenta Archaeologica 14, University of California.

Bosma, A. A., N. A. de Haan, G. J. A. Arkesteijn, F. Yang, M. Yerle, and C. Zijlstra. 2004. Comparative Chromosome Painting Between the Domestic Pig (Sus scrofa) and Two Species of Peccary, the Collared Peccary (Tayassu tajacu) and the White-lipped Peccary (T.pecari): A Phylogenetic Perspective. *Cytogenetic and Genome Research* 105:115-121.

Bottema, S. 1994. "The Prehistoric Environment of Greece: A Review of the Palynological Record," in *Beyond the Site: Regional Studies in the Aegean Area*. Edited by P. N. Kardulias, pp. 45-68. London: University Press of America.

Boyd-Dawkins, W. 1902. Remains of Animals Found in the Dictean Cave in 1901. *Man* 114:162-165.

Brain, C. K. 1969. The Contribution of Namib Desert Hottentots to an Understanding of Australopithecine Bone Accumulations. *Scientific Papers of the Namib Research Station* 39:13-22.

Bromage, T. G., Boyde, A. 1984. Microscopic Criteria for the Determination of Directionality of Cutmarks on Bone. *American Journal of Physical Anthropology* 65:359-366.

Brothwell, D., and E. Higgs. 1970. *Science In Archaeology: A Survey of Progress and Research*. New York: Praeger Publishers.

Brumfiel, E. M., and T. K. Earle. Editors. 1987. *Specialization, Exchange, and Complex Societies. Cambridge*: Cambridge University Press.

Bull, G., and S. Payne. 1982. "Tooth Eruption and Epiphyseal Fusion in Pigs and Wild Boar," in *Ageing and Sexing Animal Bones from Archaeological Sites*, vol. 109, British Series. Edited by B. Wilson, C. Grigson, and S. Payne, pp. 55-72. Oxford: BAR.

Bunn, H. T., L. E. Bartram, and E. M. Kroll. 1988. Variability in bone assemblage formation from Hadza hunting, scavenging, and carcass processing. *Journal of Anthropological Archaeology* 7:412-457.

Bunn, H. T., and E. M. Kroll. 1986. Systematic Butchery by Plio/Pleistocene Hominids at Olduvai Gorge, Tanzania. *Current Anthropology* 27:431-452.

Capaldo, S. D., and R. J. Blumenshine. 1994. A Quantitative Diagnosis of Notches Made By Hammerstone Percussion and Carnivore Gnawing on Bovid Longbones. *American Antiquity* 59:724-748.

Caskey, J. L. 1954. Excavations at Lerna, 1952-1953. *Hesperia* XXIII:3-30.

—. 1955. Excavations at Lerna, 1954. *Hesperia* XXIV:25-49.

Chadwick, J. 1967. *The Decipherment of Linear B*. Cambridge.

Chang, C. 1994. "Sheep for the Ancestors: Ethnoarchaeology and the Study of Ancient Pastoralism," in *Beyond the Site: Regional Studies in the Aegean Area*. Edited by P. N. Kardulias, pp. 353-371. London: University Press of America.

Chapman, R. 1996. ""Inventiveness and Ingenuity"? Craft Specialization, Metallurgy, and the West Mediterranean Bronze Age," in *Craft Specialization and Social Evolution*, vol. 6. Edited by B. Wailes, pp. 73-83. Philadelphia: University of Pennsylvania Museum of Archaeology and Anthropology.

Cherry, J. F., J. L. Davis, A. Demitrack, E. Mantzourani, T. F. Strasser, and L. E. Talalay. 1988. Archaeological Survey in an Artifact-Rich Landscape: A Middle Neolithic Example from Nemea, Greece. *American Journal of Archaeology* 92:159-176.

Childe, V. G. 1930. *The Bronze Age*. Cambridge: Cambridge University Press.

—. 1958. *The Prehistory of European Society*. Harmondsworth: Penguin.

—. 1963. *Social Evolution*. London: Collins.

Choquenot, D., and W. A. Ruscoe. 2003. Landscape Complementation and Food Limitation of Large Herbivores: Habitat-related Constraints on the Foraging Efficiency of Wild Pigs. *Journal of Animal Ecology* 72:14-26.

Clark, G. 1987. Faunal Remains and Economic Complexity. *Archaeozoologia* I:183-194.

Clason, A. T. Editor. 1975. *Archaeozoological Studies*. Amsterdam: Elsevier.

Clason, A. T., and W. Prummel. 1977. Collecting, Sieving, and Archaeozoological Research. *Journal of Archaeological Science* 4:171-175.

Clutton-Brock, J. 1978. "Bones for the Zoologist," in *Approaches to Faunal Analysis in the Middle East*, vol. 2, Peabody Museum Bulletin 2. Edited by R. H. Meadow and M. A. Zeder, pp. 49-51. Cambridge, MA: Peabody Museum, Harvard University.

—. 1982. "The Animal Bones," in *Excavations in Chios 1938-1955*. Edited by S. Hood, pp. 678-697. London: British School of Archaeology at Athens.

Clutton-Brock, J., and C. Grigson. Editors. 1983. *Animals and Archaeology*. Vol. 163: BAR International Series.

Cohen, R., and E. R. Service. Editors. 1978. *Origins of the State: The Anthropology of Political Evolution*. Philadelphia: Institute for the study of human issues.

Coleman, J. E. Editor. 1977. *Keos. Volume I. Kephala. A Late Neolithic Settlement and Cemetary*. Princeton: American School of Classical Studies at Athens.

Cosmopoulos, M. B. 1991. *The Early Bronze 2 in the Aegean*. Vol. XCVIII. Studies in Mediterranean Archaeology. Jonsered: Paul Astroms Forlag.

Cosmopoulos, M. B., H. J. Greenfield, and D. Ruscillo. 2003. "Animal and Marine Remains from the New Excavations at Eleusis: An Interim Report," in *Zooarchaeology in Greece: Recent Advances, vol. British School at Athens Studies* 9. Edited by E. Kotjabopoulou, Y. Hamilakis, P. Halstead, C. Gamble, and P.

Elefanti, pp. 145-152. London: The Managing Committee, British School at Athens.

Coy, J. 1977. "Animal Remains," in *Keos. Volume I. Kephala. A Late Neolithic Settlement and Cemetary*. Edited by J. E. Coleman, pp. 129-132. Princeton: American School of Classical Studies at Athens.

Crabtree, P. J. 1990. "Zooarchaeology and Complex Societies: Some Uses of Faunal Analysis for the Study of Trade, Social Status, and Ethnicity," in *Archaeological Method and Theory*: Volume 2. Edited by M. B. Schiffer, pp. 155-205. Tucson: University of Arizona Press.

Curci, A., and A. Tagliacozzo. 2003. "Economic and Ecological Evidence from the Vertebrate Remains of the Neolithic Site of Makri (Thrace-Greece)," in *Zooarchaeology in Greece: Recent Advances, British School at Athens Studies* 9. Edited by E. Kotjabopoulou, Y. Hamilakis, P. Halstead, C. Gamble, and P. Elefanti, pp. 123-131. London: The British School at Athens.

Day, L. P. 1984. Dog Burials in the Greek World. *American Journal of Archaeology* 88:21-32.

Deniz, E., and S. Payne. 1982. "Eruption and Wear in the Mandibular Dentition as a Guide to Ageing Turkish Angora Goats," in *Ageing and Sexing Animal Bones from Archaeological Sites*, vol. 109, British Series. Edited by B. Wilson, C. Grigson, and S. Payne, pp. 155-205. Oxford: British Archaeological Reports.

Diamond, J. 1995. "Why is a Cow Like a Pyramid?" in *Natural History*, vol. 95, pp. 10-12,74-77.

—. 1997. *Guns, Germs, and Steel*. New York: W.W. Norton and Company.

—. 2002. Evolution, Consequences and Future of Plant and Animal Domestication. *Nature* 418:700-707.

Dickinson, O. 1994. *The Aegean Bronze Age*. Cambridge World Archaeology. Cambridge: Cambridge University Press.

Diener, P., and E. E. Robkin. 1978. Ecology, Evolution, and the Search for Cultural Origins: The Question of Islamic Pig Prohibition. *Current Anthropology* 19:493-540.

Dobney, K. 1999. Interpreting Developmental Stress in Archaeological Pigs: The Chronology of Linear Enamel Hypoplasia. *Journal of Archaeological Science* 27:597-607.

Dobson, M. 1998. Mammal Distributions in the Western Mediterranean: The Role of Human Intervention. *Mammal Review* 28:77-88.

Drew, K. R. 1976. The Farming of Red Deer in New Zealand. *World Review of Animal Production* 12:49-60.

Ducos, P. 1978. ""Domestication" Defined and Methodological Approaches to its Recognition in Faunal Assemblages," in *Approaches to Faunal Analysis in the Middle East*, Peabody Museum Bulletins 2. Edited by R. H. Meadow and M. A. Zeder, pp. 53-68. Cambridge, MA: Peabody Museum, Harvard University.

Edwards, C. J., D. E. MacHugh, and K. M. Dobney. 2004. Ancient DNA Analysis of 101 Cattle Remains: Limits and Prospects. *Journal of Archaeological Science* 31:695-710.

Eisenberg, J. F. 1981. *The Mammalian Radiations: An Analysis of Trends in Evolution, Adaptation, and Behavior*. Chicago: The University of Chicago Press.

Evans, A. 1921. *The Palace of Minos at Knossos I.* London: Macmillan.

Evans, J. D., and C. Renfrew. 1968. *Excavations at Saliagos Near Antiparos*. London: British School of Archaeology at Athens.

Evans, R. K. 1978. "Early Craft Specialization: An Example from the Balkan Chalcolithic," in *Social Archaeology: Beyond Subsistence and Dating*. Edited by Redman. New York.

Fillios, M. 2006. *Measuring Complexity in Early Bronze Age Greece*. Minneapolis, MN: Unpublished Doctoral Dissertation, Department of Anthropology, UMN.

Fitton, J. L. 1996. *The Discovery of the Greek Bronze Age*. Cambridge, MA: Harvard University Press.

Flannery, K. V. 1968. "Archaeological Systems Theory and Early America," in *Anthropological Archaeology in the Americas*. Edited by B. J. Meggers, pp. 67- 87. Washington, D.C.

—. 1972. The Cultural Evolution of Civilization. *Annual Review of Ecology and Systematics* 3:339-426.

—. 1983. "Early Pig Domestication in the Fertile Crescent: A Retrospective Look," in *The Hilly Flanks and Beyond: Essays on the Prehistory of Southwestern Asia. Presented to Robert J. Braidwood November 15, 1982*, Studies in Ancient Oriental Civilization No.36. Edited by C. Young, T., P. Smith, and P. Mortensen, pp. 163-188. Chicago: The Oriental Institute of the University of Chicago.

Forsen, J. 1992. *The Twilight of the Early Helladics: A Study of the Disturbances in East-Central and Southern Greece Towards the End of the Early Bronze Age.* Jonsered: Paul Astroms Forlag.

French, E. B., and K. A. Wardle. Editors. 1986. *Problems in Greek Prehistory. Centenary Conference of the British School of Archaeology at Athens*. Manchester: Bristol Classical Press.

Friedl, H. 1984. *Tierknochenfunde aus Kassope/Griechenland* (4.-1. Jh. v. Chr.), Universitat Munchen.

Galvin, K. F. 1987. "Forms of finance and forms of production: the evolution of specialized livestock production in the ancient Near East," in *Specialization,Exchange, and Complex Societies*. Edited by E. M. Brumfiel and T. K. Earle, pp.119-129. Cambridge: Cambridge University Press.

Gamble, C. 1978. "The Bronze Age Animal Economy from Akrotiri: A Preliminary Analysis," in *Thera and the Aegean World I*. Edited by C. Doumas, pp. 745-753. London.

—. 1979. "Surplus and Self-Sufficiency in the Cycladic Subsistence Economy," in *Papers in Cycladic Prehistory*, Monograph XIV, Institute of Archaeology. Edited by J. L. Davis and J. F. Cherry, pp. 122-134. Los Angeles: University of California.

—. 1980. "Animal Husbandry, Population and Urbanisation," in *An Island Polity: The Archaeology of Exploitation on Melos*. Edited by C. Renfrew, Wagstaff, J.M., pp. 161-171. Cambridge: Cambridge University Press.

—. 1981. "Social Control and the Economy," in *Economic Archaeology*, vol. 96. Edited by A. Sheridan and G. Bailey, pp. 215-229: BAR.

—. 1982. "Animal Husbandry, Population and Urbanisation," in *An Island Polity*. Edited by C. Renfrew and M. Wagstaff, pp. 161-180. Cambridge: Cambridge University Press.

—. 1997. "The Animal Bones from Klithi," in *Klithi: Paleolithic Settlement and Quaternary Landscapes in Northwest Greece, vol. 1: Excavation and intra-site analysis at Klithi*. Edited by G. N. Bailey, pp. 207-244. Cambridge: McDonald Institute for Archaeological Research.

—. 1999. "Faunal Exploitation at Klithi: A Late Glacial Rockshelter in Epirus, Northwestern Greece," in *The Paleolithic Archaeology of Greece and Adjacent Areas*. British School at Athens Studies 3. Edited by G. N. Bailey, E. Adam, E. Panagopoulou, C. Perles, and K. Zachos, pp. 179-187. London: The Managing Committee, British School at Athens.

Gamble, C., and G. Bailey. 1994. "The Faunal Specialist as Excavator: the impact of recovery techniques on faunal interpretation at Klithi," in *Whither Environmental Archaeology?* vol. 38, Oxbow Monograph. Edited by R. Luff and P. Rowley-Conwy, pp. 81-90. Oxford: Oxbow Books.

Gejvall, N. 1969. *Lerna: A Preclassical Site in the Argolid. Vol. I: The Fauna*. Princeton: American School of Classical Studies at Athens.

—. 1983. "Animal Bones from the Acropolis," in *The Cuirass Tomb and Other Finds at Dendra, Part 2*. Studies in Mediterranean Archaeology. Edited by P. Astrom, pp. 51-54. Goteborg: Paul Astroms Forlag.

Genov, P. V. 1999. A Review of the Cranial Characteristics of the Wild Boar (Sus scrofa Linnaeus 1758), with Systematic Conclusions. *Mammal Review* 29:205-238.

Gifford-Gonzalez, D. 1989. "Ethnographic Analogues for Interpreting Modified Bones: Some Cases from East Africa," in *Bone Modification*. Edited by R. Bonnichsen and M. H. Sorg, pp. 179-246. Orono, Maine: Center for the Study of the First Americans.

—. 2005. "Pastoralism and its Consequences," in *African Archaeology: A Critical Introduction*. Edited by A. B. Stahl, pp.? London: Blackwell Publishing.

Gilman, A. 1996. "Craft Specialization in Late Prehistoric Mediterranean Europe," in *Craft Specialization and Social Evolution*, vol. 6. Edited by B. Wailes, pp. 67-71. Philadelphia: The University of Pennsylvania Museum of Archaeology and Anthropology.

Gimbutas, M., S. Winn, and D. Shimabuku. Editors. 1989. *Achilleion, A Neolithic Settlement in Thessaly, Greece, 6400-5600 B.C.* Monumenta Archaeologica 14. Los Angeles: Institute of Archaeology, University of California.

Gould, R. A. 1969. Subsistence Behavior Among the Western Desert Aborigines of Australia. *Oceania* 39:253-274.

Goulding, M. J. 2001. Possible Genetic Sources of Free-living Wild Boar (Sus scrofa) in Southern England. *Mammal Review* 31:245-248.

Grant, A. 1982. "The Use of Tooth Wear as a Guide to the Age of Domestic Ungulates," in *Ageing and Sexing Animal Bones from Archaeological Sites*, vol. 109, BAR British Series. Edited by B. Wilson, C. Grigson, and S. Payne, pp. 91-108. Oxford: British Archaeological Reports.

Greenfield, H. J. 1999. The Origins Of Metallurgy: Distinguishing Stone From Metal Cut-Marks On Bones From Archaeological Sites. *Journal of Archaeological Science* 26:797-808.

Greenfield, H. J., and K. Fowler. 2003. "Megalo Nisi Galanis and the Secondary Products Revolution in Macedonia," in *Zooarchaeology in Greece: Recent Advances, British School at Athens Studies* 9. Edited by E. Kotjabopoulou, Y. Hamilakis, P. Halstead, C. Gamble, and P. Elefanti, pp. 133-152. London: The British School at Athens.

Greenfield, H. J., and K. D. Fowler. 2005. *The Secondary Products Revolution in Macedonia: The Zooarchaeological Remains from Megalo Nisi Galanis, a Late Neolithic-Early Bronze Age Site in Greek Macedonia*. Vol. 1414. BAR International Series 1414. Oxford: British Archaeological Reports.

Grigson, C. 1982. "Sex and Age Determination of Some Bones and Teeth of Domestic Cattle: A Review of the literature," in *Ageing and Sexing Animal Bones from Archaeological Sites,* vol. 109, British Series. Edited by B. Wilson, C. Grigson, and S.

Payne, pp. 7-23. Oxford: British Archaeological Reports.

—. 1987. "Shiqmim: Pastoralism and Other Aspects of the Animal Management in the Chalcolithic of the Northern Negev," in *Shiqmim I. Studies Concerning Chalcolithic Societies in the Northern Negev Desert, Israel (1982-1984)*, BAR International Series 356(i). Edited by T. E. Levy, pp. 219-241. Oxford: BAR.

—. 1995. "Plough and Pasture in the Early Economy of the Southern Levant," in *The Archaeology of Society in the Holy Land*. Edited by T. E. Levy, pp. 246-268. New York: Facts on File.

Grigson, C., and J. Clutton-Brock. Editors. 1984. *Animals and Archaeology: 4. Husbandry in Europe*. Vol. 227. BAR International Series. Oxford: B.A.R.

Groves, C. 1981. Ancestors For The Pigs: Taxonomy And Phylogeny of the Genus *Sus*. *Australian National University Technical Bulletin* 3:1-96.

Guiffra, E., J. M. H. Kijas, V. Amarger, O. Carlborg, J. T. Jeon, and L. Andersson. 2000. The Origin of the Domestic Pig: Independent Domestication and Subsequent Introgression. *Genetics* 154:1785-1791.

Gustafsson, M., P. Jensen, F. H. De Jonge, G. Illmann, and M. Spinka. 1999. Maternal Behaviour of Domestic Sows and Crosses Between Domestic Sows and Wild Boar. *Applied Animal Behaviour Science* 65:29-42.

Gustafsson, M., Jensen, P., De Jonge, F.H., Schuurman, T. 1999. Domestication Effects on Foraging Strategies in Pigs (Sus scrofa). *Applied Animal Behaviour Science* 62:305-317.

Haber, A., and T. Dayan. 2004. Analyzing the Process of Domestication: Hagoshrim as a Case Study. *Journal of Archaeological Science* 31:1597-1601.

Hägg, R., and D. Konsola. 1986a. "Conclusions and Prospects," in *Early Helladic Architecture and Urbanization*. Edited by R. Hägg and D. Konsola, pp. 95-101. Goteborg: Paul Astroms Forlag.

—. Editors. 1986b. *Early Helladic Architecture and Urbanization*. Vol. LXXVI. Studies in Mediterranean Archaeology. Goteborg: Paul Astroms Forlag.

Halstead, P. 1981a. "Counting Sheep in Neolithic and Bronze Age Greece," in *Pattern of the Past: Studies in Honour of David Clarke*. Edited by I. Hodder, Isaac, G., Hammond, N., pp. 307-339. Cambridge: Cambridge University Press.

—. 1981b. "From Determinism to Uncertainty: Social Storage and the Rise of the Minoan Palace," in *Economic Archaeology*, vol. 96. Edited by A. Sheridan and G. Bailey, pp. 187-213: BAR.

—. 1985. A Study of the Mandibular Teeth from Romano-British Contexts at Maxey. *East Anglian Archaeology* 27:219-224.

—. 1987. Man and Other Animals in Later Greek Prehistory. *The Annual of the British School at Athens* 82:71-83.

—. 1989. "Like Rising Damp? An Ecological Approach to the Spread of Farming in South East and Central Europe," in *The Beginnings of Agriculture*, vol. 496, BAR International Series. Edited by A. Milles, Williams, D., Gardner, N., pp. 23-53. Oxford: B.A.R.

—. 1990. "From Reciprocity to Redistribution: Modelling the Exchange of Livestock in Neolithic Greece." *Actes du 3eme Colloque International de l'Homme and l'Animal*, Oxford, 1990, pp. 19-30 16.

—. 1992. Dimini and the 'DMP': Faunal Remains and Animal Exploitation in Late Neolithic Thessaly. *The Annual of the British School at Athens* 87:29-59.

—. 1996. Pastoralism or Household Herding? Problems of Scale and Specialization in Early Greek Animal Husbandry. *World Archaeology* 28:20-42.

—. 2003. "Texts and Bones: Contrasting Linear B and Archaeozoological Evidence for Animal Exploitation in Mycenaean Southern Greece," in *Zooarchaeology in Greece: Recent Advances, British School at Athens Studies* 9. Edited by E. Kotjabopoulou, Y. Hamilakis, P. Halstead, C. Gamble, and P. Elefanti, pp. 257-261. London: The British School at Athens.

Halstead, P., Barret, J.C. Editor. 2004. *Food, Cuisine and Society in Prehistoric Greece*. Sheffield Studies in Aegean Archaeology. Oxford: Oxbow Books.

Halstead, P., and V. Isaakidou. 2004. "Faunal Evidence for Feasting: Burnt Offerings from the Palace of Nestor at Pylos," in *Food, Cuisine and Society in Prehistoric Greece*. Edited by P. Halstead and J. C. Barrett, pp. 136-154. Oxford: Oxbow Books.

Halstead, P., and G. Jones. 1980a. Early Neolithic Economy in Thessaly: Some Evidence from Excavations at Prodromos. *Anthropologika* 1:93-117.

—. 1980b. Plant and Animal Utilization at Assiros Toumba, Central Macedonia: A Preliminary Report on the Bronze Age Material. *Bulletin of the Institute of Classical Studies* 27:134.

Hamilakis, Y. 2003. "The Sacred Geography of Hunting: Wild Animals, Social Power and Gender in Early Farming Societies," in *Zooarchaeology in Greece: Recent Advances, British School at Athens Studies* 9. Edited by E. Kotjabopoulou, Y. Hamilakis, P. Halstead, C. Gamble, and P. Elefanti, pp. 239-247. London: The British School at Athens.

Hamilakis, Y., and E. Konsolaki. 2004. Pigs for the Gods: Burnt Animal Sacrifices as Embodied Rituals at a Mycenaean Sanctuary. *Oxford Journal of Archaeology* 23:135-151.

Hansen, J. M. 1985. "Paleoethnobotany in Greece. Past, Present and Future," in *Contributions to Aegean Archaeology: Studies in Honor of William A. McDonald.* Edited by N. C. Wilkie and W. D. E. Coulson, pp. 171-181. Minneapolis: Center for Ancient Studies, University of Minnesota.

—. 1988. Agriculture in the Prehistoric Aegean: Data Versus Speculation. American *Journal of Archaeology* 92:39-52.

—. 1994. "Paleoethnobotany in Regional Perspective," in *Beyond the Site: Regional Studies in the Aegean Area.* Edited by P. N. Kardulias, pp. 173-190. London: University Press of America.

Harris, M. 1974. *Cows, Pigs, Wars and Witches: The Riddles of Culture.* New York: Random House.

—. 1985. *Good to Eat: Riddle of Food and Culture.* New York: Simon and Schuster.

Hecker, H. M. 1982a. Domestication Revisited: Its Implications for Faunal Analysis. *Journal of Field Archaeology* 9:217-236.

Hecker, R. T. 1982b. A Zooarchaeological Inquiry into Pork Consumption in Egypt from Predynastic to New Kingdom Times. *Journal of the American Research Center in Egypt* 10:59-69.

Hellstrom, P. Editor. 1987. *Paradeisos: A Late Neolithic Settlement in Aegean Thrace.* Memoir 7. Stockholm: Medelhavsmuseet.

Hennig, W. 1965. *Phylogenetic Systematics.* Urbana: University of Illinois Press.

Herre, W. 1970. "The Science and History of Domestic Animals," in *Science In Archaeology.* Edited by D. Brothwell and E. Higgs, pp. 257-272. New York: Praeger Publishers.

Herzog, Z. 1997. *Archaeology of the City: Urban Planning in Ancient Israel and Its Social Implications.* Sonia and Marco Nadler Institute of Archaeology Monography Series No.13. Tel Aviv: Emery and Clair Yass Archaeological Press, Tel Aviv University.

Hesse, B. 1995. "Husbandry, Dietary Taboos and the Bones of the Ancient Near East: Zooarchaeology in the Post-Processual World," in *Methods in the Mediterranean: Historical and Archaeological Views on Texts and Archaeology*, MNEMOSYNE Bibliotheca Classica Batava. Edited by D. B. Small, pp. 197-232. New York: E.J. Brill.

Hesse, B., and P. Wapnish. 1985. *Animal Bone Archaeology: From Objectives to Analysis.* Manuals on Archaeology 5. Washington, D.C.: Taraxacum, Inc.

—. 1997. "Can Pig Remains Be Used for Ethnic Diagnosis in the Ancient Near East?" in *The Archaeology of Israel: Constructing the Past, Interpreting the Present, Journal for the Study of the Old Testament Supplement Series 237*. Edited by N. S. Asher and D. Small, pp. 238-270. Sheffield: Sheffield Academic Press LTD.

—. 1998. "Pig Use and Abuse in the Ancient Levant: Ethnoreligious Boundary-Building with Swine," in *Ancestors for the Pigs: Pigs in Prehistory*, vol. 15, MASCA Research Papers in Science and Archaeology. Edited by S. M. Nelson, pp. 123-136. Philadelphia: MASCA.

Higgs, E. S. 1962. "Fauna," in *Excavations at the Early Neolithic Site at Nea Nikomedeia, Greek Macedonia (1961 Season)*, vol. 28. Edited by R. J. Rodden, pp. 267-288: Proceedings of the Prehistoric Society.

—. Editor. 1972. *Papers in Economic Prehistory*. Cambridge: Cambridge University Press.

Higgs, E. S., I. M. Clegg, and I. A. Kinnes. 1968. "Saliagos Animal Bones," in *Excavations at Saliagos Near Antiparos*. Edited by J. D. Evans and C. Renfrew, pp. 114-117. London: British School of Archaeology at Athens.

Higgs, E. S., Vita-Finzi, C., Harris, D.R., Fagg, A.E. 1967. The Climate, Environment and Industries of Stone Age Greece, Part III. *Proceedings of the Prehistoric Society* 33:1-29.

Higham, C. F. W. 1967. Stock Rearing as a Cultural Factor in Prehistoric Europe. *Proceedings of the Prehistoric Society* 33:84-106.

Hill, K., and H. Kaplan. 1999. Life History Traits in Humans: Theory and Empirical Studies. *Annual Review of Anthropology* 28:397-430.

Hockett, B., and J. Haws. 2003. Nutritional Ecology and Diachronic Trends in Paleolithic Diet and Health. *Evolutionary Anthropology* 12:211-216.

Hodder, I., G. Isaac, and N. Hammond. Editors. 1981. *Pattern of the Past: Studies in Honour of David Clarke*. Cambridge: Cambridge University Press.

Hongo, H., and R. H. Meadow. 1998. "Pig Exploitation at Neolithic Cayonu Tepesi (Southeastern Anatolia)," in *Ancestors for the Pigs: Pigs in Prehistory*, vol. 15, MASCA Research Papers in Science and Archaeology. Edited by S. M. Nelson, pp. 77-98. Philadelphia: MASCA.

Horwitz, L. K. 1996. The Impact of Animal Domestication on Species Richness: A Pilot Study from the Neolithic of the Southern Levant. *Archaeozoologia* 8:53-70.

Hourmouziadis, G. 1979. *To Neolithiko Dimini*. Volos, Greece.

IJzereef, G. F. 1988. Animal Bones and Social Stratification. A Preliminary Analysis of the Faunal Remains from Cess-Pits in Amsterdam (1600-1850 A.D.). *Archaeozoologia* II:283-292.

Ioannidou, E. 2003. Taphonomy Of Animal Bones: Species, Sex, Age and Breed Variability of Sheep, Cattle, and Pig Bone Density. *Journal of Archaeological Science* 30:355-365.

Jameson, M. H. 1988. "Sacrifice and Animal Husbandry in Classical Greece," in *Pastoral Economies in Classical Antiquity*, Supplemental Volume 14. Edited by C. R. Whittaker, pp. 87-119. Cambridge: Cambridge Philological Society.

Jameson, M. H., C. N. Runnels, and T. H. van Andel. 1994. *A Greek Countryside: The Southern Argolid from Prehistory to the Present Day*. Stanford, C.A.: Stanford University Press.

Jarman, M. R., and H. N. Jarman. 1968. The Fauna and Economy of Early Neolithic Knossos. *The Annual of the British School at Athens* 63:239-264.

Joffe, A., H. 2004. Athens and Jerusalem in the Third Millennium: Culture, Comparison, and the Evolution of Social Complexity. *Journal of Mediterranean Archaeology* 17:247-267.

Jones, G. 1987. Agricultural Practice in Greek Prehistory. *The Annual of the British School at Athens* 82:115-123.

Jones, M., and T. Brown. 2000. Agricultural Origins: The Evidence of Modern and Ancient DNA. *The Holocene* 10:769-776.

Jordan, B. 1975. *Tierknochenfunds aus der Magula Pevkakia in Thessalien*, Universitat Munchen.

Jullien, R. 1973. Grotte de Kitsos III: La Faune: Mammifieres, Reptiles et Batraciens. *Bulletin de Correspondance Hellenique* 97:434-443.

Kapetanios, A. 2003. "The 'Socialization' of Animals in Epirus, Ikaria and Crete: the Material and Symbolic Role of Man-Animal Relations of Production in the Process of Social Formation," in *Zooarchaeology in Greece: Recent Advances, British School at Athens Studies* 9. Edited by E. Kotjabopoulou, Hamilakis,Y., Halstead, P., Gamble, C., Elefanti, P., pp. 283-290. London: The British School at Athens.

Karali, L. 2000. "Evolution Actuelle de l'Archaeozoologie en Grece dans le Neolithique et l'Age du Bronze," in *Archaeozoology of the Near East IVB*, Publicatie 32, Proceedings of the fourth international symposium on the archaeozoology of southwestern Asia and adjacent areas. Edited by M. Mashkour, A. M. Choyke, H. Buitenhuis, and F. Poplin, pp. 187-196. Groningen: ARC.

Kardulias, P. N. 1992. The Ecology of Bronze Age Flaked Stone Tool Production in Southern Greece: Evidence from Agios Stephanos and the Southern Argolid. *American Journal of Archaeology* 96:421-442.

—. Editor. 1994a. *Beyond the Site: Regional Studies in the Aegean Area.* New York: University Press of America.

—. 1994b. "Paradigms of the Past in Greek Archaeology," in *Beyond the Site: Regional Studies in the Aegean Area.* Edited by P. N. Kardulias, pp. 1-23. London: University Press of America.

Karkanas, P., N. Kyparissi-Apostolika, O. Bar-Yosef, and S. Weiner. 1999. Mineral Assemblages in Theopetra, Greece: A Framework for Understanding Diagenesis in a Prehistoric Cave. *Journal of Archaeological Science* 26:1171-1180.

Kelm, H. 1938. Die postembryonale Schaedelentwicklung des Wild und Berkshire-Schweins. *Zeitschrift fur Anatomie Und. Entwicklungsgeschichte* 108:499-559.

Kilian, K. 1986. "The Circular Building at Tiryns," in *Early Helladic Architecture and Urbanization.* Edited by R. Hägg and D. Konsola, pp. 65-71. Goteborg: Paul Astroms Forlag.

Klein, R. G., and K. Cruz-Uribe. 1984. *The Analysis of Animal Bones from Archaeological Sites.* Chicago: University of Chicago Press.

Konsola, D. 1986. "Stages of Urban Transformation in the Early Helladic Period," in *Early Helladic Architecture and Urbanization.* Edited by R. Hägg and D. Konsola, pp. 9-19. Goteborg: Paul Astroms Forlag.

Kotjabopoulou, E., Gamble, C. 2004. "Subsistence Through the Ages: Introduction," in *Zooarchaeology in Greece: Recent Advances*, British School at Athens Studies 9. Edited by E. Kotjabopoulou, Hamilakis,Y., Halstead, P., Gamble, C., Elefanti, P., pp. 111-113. London: The British School at Athens.

Kotjabopoulou, E., Y. Hamilakis, P. Halstead, C. Gamble, and P. Elefanti. Editors. 2003. *Zooarchaeology in Greece: Recent Advances.* Vol. 9. British School at Athens Studies. London: The British School at Athens.

Koufos, G. 1981. "To Osteologiko Iliko. Paratima II," in *Anaskafi se Oikismo tis Epochis Chalcou (Proimis) stin Pentapoli tou Nomou Serron.* Edited by D. Grammenos, pp. 113-115: Archaiologiki Ephemeris.

Koumouzelis, M., B. Ginter, K. Kozlowski, M. Pawlikowski, O. Bar-Yosef, R. M. Albert, M. Litynska-Zajac, E. Stworzewicz, P. Wojtal, G. Lipecki, T. Tomek, Z. Bochenski, and A. Pazdur. 2001. The Early Upper Paleolithic in Greece: The Excavations in Klisoura Cave. *Journal of Archaeological Science* 28:515-539.

Kretzoi, M., and N. A. Poulianos. 1981. Remarks on the Middle and Lower Pleistocene Vertebrate Fauna in the Petralona Cave. *Anthropos* 8:57-72.

Kunzl, C., S. Kaiser, E. Meier, and N. Sachser. 2003. Is a Wild Mammal Kept and Reared in Captivity Still a Wild Animal? *Hormones and Behavior* 43:187-196.

Kunzl, C., Sachser, N. 1999. The Behavioral Endocrinology of Domestication: A Comparison Between the Domestic Guinea Pig (Cavia aperea f. porcellus) and Its Wild Ancestor, the Cavy (Cavia aperea). *Hormones and Behavior* 35:28-37.

Lam, Y., and O. M. Pearson. 2004. The Fallibility of Bone Density Values and their Use in Archaeological Analyses. *Journal of Taphonomy* 2:99-116.

Lam, Y. M., X. Chen, and O. M. Pearson. 1999. Intertaxonomic Variability in Patterns of Bone Density and the Differential Representation of Bovid, Cervid, and Equid Elements in the Archaeological Record. *American Antiquity* 64:343-362.

Larje, R. 1987. "Animal Bones," in *Paradeisos: A Late Neolithic Settlement in Aegean Thrace*, Memoir 7. Edited by P. Hellstrom, pp. 89-118. Stockholm: Medelhavsmuseet.

Lauwerier, R. C. G. M. 1983. Pigs, Piglets and Determining the Season of Slaughtering. *Journal of Archaeological Science* 10:483-488.

Le Blay, J.-C. 1993. Hasard, Taphonomie ou Redevance? Le Cas du Porc Medieval. *Anthropozoologica* 17:7-10.

Levy, T. E. Editor. 1987. *Shiqmim I. Studies Concerning Chalcolithic Societies in the Northern Negev Desert, Israel (1982-1984).* Oxford: BAR International Series 356(i).

Lobban, R. A. 1998. "Pigs in Ancient Egypt," in *Ancestors for the Pigs: Pigs in Prehistory*, vol. 15, MASCA Research Papers in Science and Archaeology. Edited by S. M. Nelson, pp. 137-148. Philadelphia: MASCA.

Loftus, R. T., O. Ertugrul, A. H. Harba, M. A. A. El-Barody, D. E. Machugh, S. D. E. Park, and D. G. Bradley. 1999. A Microsatellite Survey of Cattle From a Centre of Origin: the Near East. *Molecular Ecology* 8:2015-2022.

Luff, R., and P. Rowley-Conwy. Editors. 1994. *Whither Environmental Archaeology?* Vol. 38. Oxbow Monograph. Oxford: Oxbow Books.

Lyman, R. L. 1992. Anatomical Consideration of Utility Curves in Zooarchaeology. *Journal of Archaeological Science* 19:7-22.

—. 1994. *Vertebrate Taphonomy*. Cambridge Manuals in Archaeology. Cambridge: Cambridge University Press.

Mace, R. 2000. Evolutionary Ecology of Human Life History. *Animal Behaviour* 59:1-10.

Mackinnon, M. 2001. High on the Hog: Linking Zooarchaeological, Literary, and Artistic Data for Pig Breeds in Roman Italy. *American Journal of Archaeology* 105:649-73.

Mainland, I. L. 2003. "Dental Microwear in Modern Greek Ovicaprids: Identifying Microwear Signatures Asssociated with a Diet of Leafy-hay," in *Zooarchaeology in Greece: Recent Advances, British School at Athens Studies* 9. Edited by E. Kotjabopoulou, Y. Hamilakis, P. Halstead, C. Gamble, and P. Elefanti. London: The Managing Committe, the British School at Athens.

Major, F. 1883. Studien zur Geschichte der Wildschweine (Gen Sus). *Zoologischer Anzeiger* VI:295-300.

Mancz, E. A. 1989. *An Examination of Changing Patterns of Animal Husbandry of the Late Bronze and Dark Ages of Nichoria in the Southwestern Peloponnese*. Doctor of Philosophy, University of Minnesota.

Marean, C. W., and C. J. Frey. 1997. Animal bones from caves to cities: Reverse utility curves as methodological artifacts. *American Antiquity* 62:698-711.

Marean, C. W., and S. Y. Kim. 1998. Mousterian Large Mammal Remains from Kobeh Cave (Zagros Mountains, Iran): Behavioral Implications for Neandertals and Early Modern Humans. *Current Anthropology* 39S:79-113.

Marean, C. W., and L. M. Spencer. 1991. Impact of Carnivore Ravaging on Zooarchaeological Measures of Element Abundance. *American Antiquity* 56:645-658.

Marshall, F., and E. Hildebrand. 2002. Cattle Before Crops: The Beginnings of Food Production in Africa. *Journal of World Prehistory* 16:99-143.

Mashkour, M., A. M. Choyke, H. Buitenhuis, and F. Poplin. Editors. 2000. *Archaeozoology of the Near East IVB*. Publicatie 32. Proceedings of the fourth international symposium on the archaeozoology of southwestern Asia and adjacent areas. Groningen: ARC.

Masseti, M. 2003. "Taxonomic and Behavioural Aspects of the Representation of Mammals in Aegean Bronze Age Art," in *Zooarchaeology in Greece: Recent Advances, British School at Athens Studies* 9. Edited by E. Kotjabopoulou, Y. Hamilakis, P. Halstead, C. Gamble, and P. Elefanti, pp. 273-282. London: The Managing Committee, the British School at Athens.

Mathews, M. K., M. C. White, and R. G. Long. 1999. Why Study the Complexity Sciences in the Social Sciences? *Human Relations* 52:439-462.

May, E., M. Reichert, and K. Hannemann. 1996. Allometric Aspects to the Determination of the Withers Height in Pigs on the Basis of the Data of M. Teichert. *Archaeozoologia* 8:125-139.

Mayer, J. J., J. M. Novak, and I. L. Brisbin. 1998. "Evaluation of Molar Size as a Basis for Distinguishing Wild Boar from Domestic Swine: Employing the Present to Decipher the Past," in *Ancestors for the Pigs: Pigs in Prehistory*, vol. 15, MASCA Research Papers in Science and Archaeology. Edited by S. M. Nelson, pp. 39-54. Philadelphia: MASCA University of Pennsylvania Museum of Archaeology and Anthropology.

McDonald, W. A., and G. R. Rapp. Editors. 1972. *The Minnesota Messenia Expedition. Reconstructing a Bronze Age Regional Environment*. Minneapolis: University of Minnesota Press.

Meadow, R. H. 1978. "Effects of Context on the Interpretation of Faunal Remains: A Case Study," in *Approaches to Faunal Analysis in the Middle East*, vol. 2, Peabody Museum Bulletin 2. Edited by R. H. Meadow and M. A. Zeder, pp. 15-21. Cambridge, MA: Peabody Museum, Harvard University.

Meadow, R. H., and M. A. Zeder. Editors. 1978. *Approaches to Faunal Analysis in the Middle East*. Peabody Museum Bulletin 2. Cambridge, MA: Peabody Museum, Harvard University.

Meggers, B. J. Editor. 1968. *Anthropological Archaeology in the Americas*. Washington, D.C.

Miracle, P., and N. Milner. Editors. 2002. *Consuming Passions and Patterns of Consumption*. McDonald Institute Monographs. Cambridge: McDonald Institute for Archaeological Research.

Moberg, K. 1992. The Animal Bones. Appendix in Excavations in the Levandis Sector at Asine, 1989. *Opuscula Atheniensia* 19:66-68.

Monahan, C. 1998. The Hadza CarcassTransport Debate Revisited and its Archaeological Implications. *Journal of Archaeological Science* 25:405-424.

Moore, A. M. T. 1979. *The Neolithic of the Levant*, Oxford University.

Morgan, L. 1983. *The Miniature Wall Paintings of Thera: A Study in Aegean Culture and Iconography*. Cambridge: Cambridge University Press.

Muir, R. J., and J. C. Driver. 2002. Scale of Analysis and Zooarchaeological Interpretation: Pueblo III Faunal Variation in the Northern San Juan Region. *Journal of Anthropological Archaeology* 21:165-199.

Mylonas, G. E. 1961. *Eleusis and the Eleusinian Mysteries*. Princeton: Princeton University Press.

Nathusius, H. V. 1864. *Vorstudien fur Geschichte und Zuecht. der Haustiere zunachst am Schweineschadel*. Vol. 25. Ex.

Nelson, S. M. Editor. 1998. *Ancestors for the Pigs: Pigs in Prehistory*. Vol. 15. MASCA Research Papers in Science and Archaeology. Philadelphia: MASCA.

Newton, S. 2003. "Theopetra Cave and the Paleolithic-Mesolithic Transition in Southern Europe," in *Zooarchaeology in Greece: Recent Advances, vol. British School at Athens Studies* 9. Edited by E. Kotjabopoulou, Y. Hamilakis, P. Halstead, C. Gamble, and P. Elefanti, pp. 115-122. London: The Managing Committe, The British School at Athens.

Nobis, G. 1993. Archaozoologische Untersuchungen von Tierresten aus dem 'Palast des Nestor' bei Pylos in Messenien, SW-Peloponnes. *Zeitschrift fur Archaologie* 27:151-173.

Novecosky, B. J., and P. R. W. Popkin. 2005. Canidae Volume Bone Mineral Density Values: An Application to Sites in Western Canada. *Journal of Archaeological Science* 32:1677-1690.

Nowak, R. M. 1999. *Walker's Mammals of the World.* Vol. II. Baltimore.

O'Connor, T. P. 1996. A Critical Overview of Archaeological Animal Bone Studies. *World Archaeology* 28:5-19.

O'Day, S. J., and E. A. Van Neer Wim. Editors. 2004. *Behaviour Behind Bones: The zooarchaeology of ritual, religion, status and identity.* Proceedings of the 9th Conference of the International Council of Archaeozoology, Durham, August 2002. Oxford: Oxbow Books.

Oksanen, T. A., I. Jokinen, E. Koskela, T. Mappes, and H. Vilpas. 2004. Manipulation of Offspring Number and Size: Benefits of Large Body Size at Birth Depend Upon the Rearing Environment. *Journal of Animal Ecology* 72:321-330.

Olsen, S. L. Editor. 1988. *The Identification of Stone and Metal Tool Marks on Bone Artifacts. Vol. 452. Scanning Electron Microscopy in Archaeology*: BAR International.

Osborne, R. 2004. Greek Archaeology: A Survey of Recent Work. *American Journal of Archaeology* 108:87-102.

Panagiotakopulu, E. 1999. An Examination of Biological Materials from Coprolites from XVIII Dynasty Amarna, Egypt. *Journal of Archaeological Science* 26:547-551.

Pappa, M., P. Halstead, K. Kotsakis, and D. Urem-Kotsou. 2004. "Evidence for Large-scale Feasting at Late Neolithic Makriyalos, N Greece," in *Food, Cuisine and Society in Prehistoric Greece*, Sheffield Studies in Aegean Archaeology. Edited by P. Halstead and J. C. Barret, pp. 16-44. Oxford: Oxbow Books.

Pavlakis, P., A. Fistani, and N. Symeonidis. 1999. "New Evidence of Middle Pleistocene Mammalian Fauna from the Perama Cave (Ioannina, Greece)," in *The Paleolithic Archaeology of Greece and Adjacent Areas.* British School at Athens Studies 3. Edited by G. N. Bailey, E. Adam, E. Panagopoulou, C. Perles, and K. Zachos, pp. 97-107. London: The Managing Committee, British School at Athens.

Payne, S. 1972. "On the Interpretation of Animal Bone from Archaeological Sites," in *Papers in Economic Prehistory*. Edited by E. S. Higgs, pp. 65-81. Cambridge: Cambridge University Press.

—. 1973. Kill-off Patterns in Sheep and Goats: The Mandibles from Asvan Kale. *Anatolian Studies* 23:281-303.

—. 1975. "Partial Recovery and Sampling Bias," in *Archaeozoological Studies*. Edited by A. T. Clason, pp. 7-17. Amsterdam: Elsevier.

—. 1982. "Faunal Evidence for Environmental/Climatic Change at Franchthi Cave (Southern Argolid, Greece), 25,000 BP to 5,000 BP - Preliminary Results," in *Paleoclimates, Paleoenvironments and Human Communities in the Eastern Mediterranean Region in Later Prehistory*, BAR. Edited by J. L. Bintliff and W. van Zeist, pp. 133-136. Oxford: BAR S133.

—. 1985. "Zoo-Archaeology in Greece: A Reader's Guide," in *Contributions to Aegean Archaeology: Studies in Honor of William A. McDonald*. Edited by N. C. Wilkie and W. D. E. Coulson, pp. 211-244. Minneapolis: Center for Ancient Studies, University of Minnesota.

Payne, S., and G. Bull. 1988. Components of Variation in Measurements of Pig Bones and Teeth, and the Use of Measurements to Distinguish Wild from Domestic Pig Remains. *Archaeozoologia* 2:27-66.

Pearce, J., and R. Luff. 1994. "The Taphonomy of Cooked Bone," in *Whither Environmental Archaeology?* vol. 38, Oxbow Monograph. Edited by R. Luff and P. Rowley-Conwy, pp. 51-56. Oxford: Oxbow Books.

Perles, C. 2001. *The Early Neolithic in Greece: The First Farming Communities in Europe*. Cambridge World Archaeology. Cambridge: Cambridge University Press.

Pernier, L. 1935. *Il Palazzo Minoico de Festos. I: Gli Strati Piu Antichi e il Primo Palazzo*: Libreria dello Stato.

Pickering, T. R. 2002. Reconsideration of criteria for differentiating faunal assemblages accumulated by hyenas and hominids. *International Journal of Osteoarchaeology* 12:127-141.

Prummel, W. 1987a. Atlas for Identification of Foetal Skeletal Elements of Cattle, Horse, Sheep and Pig: Part.1. *Archaeozoologia* 1:23-30.

—. 1987b. Atlas for Identification of Foetal Skeletal Elements of Cattle, Horse, Sheep and Pig: Part.2. *Archaeozoologia* 1:11-42.

—. 1988. Atlas for Identification of Foetal Skeletal Elements of Cattle, Horse, Sheep and Pig: Part.3. *Archaeozoologia* 2:13-26.

—. 1989. Appendix to Atlas for Identification of Foetal Skeletal Elements of Cattle, Horse, Sheep and Pig. *Archaeozoologia* 3:71-78.

Prummel, W., and H. Frisch. 1986. A Guide for the Distinction of Species, Sex and Body Side in Bones of Sheep and Goat. *Journal of Archaeological Science* 13:567-577.

Pullen, D. J. 1986. "The Early Bronze Age Settlement on Tsoungiza Hill, Ancient Nemea," in *Early Helladic Architecture and Urbanization*, vol. 76, Studies in Mediterranean Archaeology. Edited by R. Hägg and D. Konsola, pp. 73-78. Goteborg.

—. 1988. The Earlier Phases of the Early Bronze Age at Tsoungiza Hill, Ancient Nemea (Abstract). *American Journal of Archaeology* 92:252.

—. 1992. Ox and Plow in the Early Bronze Age Aegean. *American Journal of Archaeology* 96:45-54.

—. 1994. "Modeling Mortuary Behavior on a Regional Scale: A Case Study from Mainland Greece in the Early Bronze Age," in *Beyond the Site: Regional Studies in the Aegean Area*. Edited by P. N. Kardulias, pp. 113-136. London: University Press of America.

—. 2003. "Site Size, Territory, and Hierarchy: Measuring Levels of Integration and Social Change in Neolithic and Bronze Age Aegean Societies." *Metron: Measuring the Aegean Bronze Age*. Proceedings of the 9th International Aegean Conference, Yale University, 2003, pp. 29-37.

—. 2006 (in press). *The Early Bronze Age Village on Tsoungiza Hill*. Princeton: American School of Classical Studies at Athens.

Radice, B. 1971. *Who's Who in the Ancient World*. New York: Penguin Books.

Rapp, G., Aschenbrenner, S.E. Editor. 1978. *Excavations at Nichoria in Southwest Greece. Vol. 1: Site, Environs, and Techniques*. Minneapolis: The University of Minnesota Press.

Rapp, G., and J. C. Kraft. 1994. "Holocene Coastal Change in Greece and Aegean Turkey," in *Beyond the Site: Regional Studies in the Aegean Area*. Edited by P. N. Kardulias, pp. 69-90. London: University Press of America.

Reader, S. M., and K. N. Laland. 1999. Do Animals Have Memes? *Journal of Memetics - Evolutionary Models of Information Transmission* 3:16-21.

Redding, R. W. 1991. The Role of the Pig in the Subsistence System of Ancient Egypt: A Parable on the Potential of Faunal Data. *Animal Use And Culture Change*:20-30.

Redding, R. W., and M. Rosenberg. 1998. "Ancestral Pigs: A New (Guinea) Model for Pig Domestication in the Middle East," in *Ancestors for the Pigs: Pigs in Prehistory*, vol. 15, MASCA Research Papers in Science and Archaeology. Edited by S. M. Nelson, pp. 65-76. Philadelphia: MASCA.

Redman, C. L. 1978. *Social Archaeology: Beyond Subsistence and Dating*. New York.

Reed, C. A. 1969. "The Pattern of Animal Domestication in the Prehistoric Near East," in *The Domestication and Exploitation of Plants and Animals*. Edited by P. J. Ucko and G. W. Dimbleby, pp. 362-380. Chicago: Aldine Publishing Company.

Reese, D. S. 1994a. "Recent Work in Greek Zooarchaeology," in *Beyond the Site: Regional Studies in the Aegean Area*. Edited by P. N. Kardulias, pp. 191-221. New York: University Press of America.

—. 1994b. "The Mammal Bones and Invertebrates from Agios Stephanos," in *Agios Stephanos, Results of Excavations in Lakonia, 1973-1977*, BSA Supplemental Volume. Edited by W. Taylour and R. Janko. London: Thames and Hudson.

Reichstein, H. 1982. Erste Ergebnisse von Untersuchungen an Tierknochen aus bronzezeitlichen Siedlungsschichten im nordlichen Griechenland (Ausgrabung Kastanas). *Jahrbuch des Romisch-Germanischen Zentralmuseums Mainz* 26:239-270.

Renard, J. 1995. *Le Peloponnese au Bronze Ancien*. Liege: Universite de Liege.

Renfrew, C. 1972. *The Emergence of Civilization: The Cyclades and the Aegean in the Third Millennium B.C.* London.

—. 1979. *Problems in European Prehistory*. Edinburgh: Edinburgh University Press.

Renfrew, C., M. Gimbutas, and E. S. Elster. 1986. *Excavations at Sitagroi: A Prehistoric Village in Northeast Greece*. Vol. 1. Monumenta Archaeologica Volume 13. Los Angeles: Regents of the University of California.

Renfrew, C., and M. Wagstaff. Editors. 1982. *An Island Polity*. Cambridge: Cambridge University Press.

Richards, M. P., P. B. Pettitt, E. Trinkaus, F. H. Smith, M. Paynovic, and I. Karavanic. 2000. Neanderthal Diet at Vindya and Neanderthal Predation: The Evidence from Stable Isotopes. *PNAS* 97.

Rienders, H. R., and W. Prummer. Editors. 2003. *Housing in New Halos: A Hellenistic Town in Thessaly, Greece*. Lisse: A.A. Balkema.

Roberts, M. R. 2001. The Place of Farm Animal Species in the New Genomics World of Reproductive Biology. *Biology of Reproduction* 64:409-417.

Rolett, B. V., and M. Chiu. 1994. Age Estimation of Prehistoric Pigs (Sus scrofa) by Molar Eruption and Attrition. *Journal of Archaeological Science* 21:377-386.

Rosen, S. A. 1997. "Craft Specialization and the Rise of Secondary Urbanism: A View from the Southern Levant," in *Urbanism in Late Antiquity: From Mesopotamia to Crete*. Edited by W. E. Aufrecht, N. A. Mirau, and S. W. Gauley, pp. 82-89. Sheffield: Sheffield Academic Press.

Runnels, C. N., and T. H. van Andel. 1987. The Evolution of Settlement in the Southern Argolid, Greece: An Economic Explanation. *Hesperia* 56:303-334.

Ruscillo, D. 2003. "Alternative Methods for Identifying Sex from Archaeological Animal Bone," in *Zooarchaeology in Greece: Recent Advances, British School at Athens Studies* 9. Edited by E. Kotjabopoulou, Y. Hamilakis, P. Halstead, C. Gamble, and P. Elefanti, pp. 37-44. London: The Managing Committe, The British School at Athens.

Russell, N. 2002. The Wild Side of Animal Domestication. *Society and Animals: Journal of Human-Animal Studies* 10:1-8.

Sampson, A. 1987. The Early Helladic Graves of Manika. *Aegaeum* 1?

Savina, M., S. E. Aschenbrenner, and N. Wilkie. 1991. Terra Incognita no Longer: Archaeological Survey in Grevana. *American Journal of Archaeology* 95:328 (abstract).

Schallin, A.-L. 1997. "Urban Centres, Central Places and Nucleation in Greek Islands Versus the Greek Mainland," in *Urbanization in the Mediterranean in the 9th to 6th Centuries B.C., Acta Hyperborea 7 Danish Studies in Classical Archaeology*. Edited by H. D. Andersen, W. Horsnaes Helle, S. Houby-Nelson, and A. Rathje, pp. 17-44. Copenhagen: Museum Tusculanum Press, University of Copenhagen.

Schiffer, M. B. Editor. 1990. *Archaeological Method and Theory*: Volume 2. Tucson: University of Arizona Press.

Schley, L., and T. J. Roper. 2003. Diet of Wild Boar Sus scrofa in Western Europe, with Particular Reference to Consumption of Agricultural Crops. *Mammal Review* 33:43-56.

Schliemann, H. 1880. *Ilios, The City and Country of the Trojans*. London: John Murray.

Schmid, E. 1965. "Die Seidi-Hohle, eine jungpalaolithische Station in Griechenland." *IV Colloque International de Spelealogie, 1963, Athens, 1965*, pp. 163-174.

Schwartz, C. A. 1982. "The Fauna from Early Neolithic Sesklo," in *The Early Neolithic I Settlement at Sesklo: An Early Farming Community in Thessaly, Greece*. Edited by M. Wijnen, pp. p.112. Leiden: Leiden University Press.

Seger, J. D. Editor. 1996. *Retrieving the Past: Essays on Archaeological Research and Methodology in Honor of Gus W. Van Beek*. Mississippi: Cobb Institute of Archaeology.

Sheridan, A., and G. Bailey. Editors. 1981. *Economic Archaeology*. Vol. 96. BAR.

Sherratt, A. 1981. "Plough and Pastoralism: Aspects of the Secondary Products Revolution," in *Pattern of the Past: Studies in Honour of David Clarke*. Edited by I. Hodder, G. Isaac, and N. Hammond, pp. 261-305. Cambridge: Cambridge University Press.

Shipman, P. 1983. "Early hominid lifestyles: Hunting and gathering or foraging and scavenging?" in *Animals and Archaeology*, vol. 163. Edited by J. Clutton-Brock and C. Grigson, pp. 31-49: BAR International Series.

Shipman, P., and J. Rose. 1983. Early Hominid Hunting, Butchering, and Carcass-Processing Behaviors: Approaches to the Fossil Record. Journal of *Anthropological Archaeology* 2:57-98.

Silver, I. A. 1969. "The Aging of Domestic Animals," in *Science in Archaeology*. Edited by D. Brothwell and E. Higgs, pp. 250-268. New York: Praeger Publishers.

Siracusano, G. 2004. "Animal husbandry and centralized cultures. How social and political factors can influence rural lifestyle.," in *Behaviour Behind Bones: The zooarchaeology of ritual, religion, status and identity*, Proceedings of the 9th Conference of the International Council of Archaeozoology, Durham, August 2002. Edited by S. J. O'Day, W. van Neer, and A. Ervynck, pp. 190-197. Oxford: Oxbow Books.

Sloan, R. E., and M. A. Duncan. 1978. "Zooarchaeology of Nichoria," in *Excavations at Nichoria in Southwest Greece, vol. 1: Site, Environs, and Techniques*. Edited by G. Rapp and S. E. Aschenbrenner, pp. 60-77. Minneapolis: The University of Minnesota Press.

Small, D. B. Editor. 1995. *Methods in the Mediterranean: Historical and Archaeological Views on Texts and Archaeology*. MNEMOSYNE Bibliotheca Classica Batava. New York: E.J. Brill.

Smith, M. L. Editor. 2003. *The Social Construction of Ancient Cities*. Washington, D.C.: Smithsonian Books.

Snodgrass, A. M. 1985. The New Archaeology and the Classical Archaeologist. *American Journal of Archaeology* 89:31-37.

Snyder, L. M., and W. E. Klippel. 2003. "From Lerna to Kastro: Further Thoughts on Dogs as Food in Ancient Greece; Perceptions, Prejudices, and Reinvestigations," in *Zooarchaeology in Greece: Recent Advances, vol. British School at Athens Studies* 9. Edited by E. Kotjabopoulou, Y. Hamilakis, P. Halstead, C. Gamble, and P. Elefanti, pp. 221-231. London: The Managing Committee, British School at Athens.

Soderberg, J. A. 2003. *Feeding Community: Urbanization, Religion, and Zooarchaeology at Clonmacnoise, an Early Medieval Irish Monastery*. Doctoral Dissertation, University of Minnesota.

Sordinas, A. 1969. Investigations of the Prehistory of Corfu During 1964-1966. *Balkan Studies* 10:393-424.

Spatz, C. 1993. *Basic Statistics: Tales of Distributions*. Pacific Grove, CA: Brooks/Cole Publishing Company.

Spinka, M., G. Illmann, F. De Jonge, M. Andersson, T. Schuurman, and P. Jensen. 2000. Dimensions of Maternal Behaviour Characteristics in Domestic and Wild X Domestic Crossbred Sows. *Applied Animal Behaviour Science* 70:99-114.

Stanzel, M. 1991. *Die Tierreste aus dem Artemis-/Apollon-Heiligtum bei Kalapodi in Bootien/Griechenland*, Ludwig-Maximilians-Universitat.

Stemp, W. J., and M. Stemp. 2001. UBM Laser Profilometry And Lithic Use-Wear Analysis: A Variable Length Scale Investigation Of Surface Topography. *Journal of Archaeological Science* 28:81-88.

Stephan, E. 2000. Oxygen Isotope Analysis of Animal Bone Phosphate: Method Refinement, Influence of Consolidants, And Reconstruction of Paleotemperatures for Holocene Sites. *Journal of Archaeological Science* 27:523-535.

Stibbe, A. 2003. As Charming as a Pig: The Discursive Construction of the Relationship Between Pigs and Humans. *Society & Animals* 11:375-392.

Sutton, R. F. 1990. The Nemea Valley Archaeological Project: A Preliminary Report. *Hesperia* 59:579-659.

Thomas, K. D. 1996. Zooarchaeology: Past, Present and Future. *World Archaeology* 28:1-4.

Tikhonov, V., A. Sruoga, and V. Bobovich. 2004. The Development of Molecular-Genetic Euro-Asian Heterozygosity Genome in Microevolution of Sus scrofa scrofa as a Start Success of the Formation of Domestic Breeds. *Acta Zoologica Lituanica* 14:3-18.

Toufexis, G. 2003. "Animals in the Neolithic Figurine Art of Thessaly," in *Zooarchaeology in Greece: Recent Advances, British School at Athens Studies* 9. Edited by E. Kotjabopoulou, Y. Hamilakis, P. Halstead, C. Gamble, and P. Elefanti, pp. 263-271. London: The Managing Committee, The British School at Athens.

Tzavella-Evjen, H. 1985. *Lithares: An Early Bronze Age Settlement in Boeotia.* Occasional Paper 15, Institute of Archaeology, UCLA. Los Angeles: Regents of the University of California.

Ucko, P. J., and G. W. Dimbleby. Editors. 1969. *The Domestication and Exploitation of Plants and Animals*. Chicago: Aldine Publishing Company.

Ungar, P. S., and F. M'Kirera. 2003. A Solution to the Worn Tooth Conundrum in Primate Functional Anatomy. *PNAS.*

Valamoti, S. M. 2004. *Plants and People in Late Neolithic and Early Bronze Age Northern Greece: An archaeobotanical investigation*. Vol. 1258. BAR International Series 1258. Oxford: BAR Publishing.

Valmin, M. N. 1938. *The Swedish Messenian Expedition: Malthi Dorion*. Lund: Gleerup.

van der Veen, M., and N. Fieller. 1982. Sampling Seeds. *Journal of Archaeological Science* 9:287-298.

Ventris, M., and J. Chadwick. 1953. *The Journal of Hellenic Studies* 73:84-103.

Vernesi, C., B. Crestanello, E. Pecchioli, D. Tartari, D. Caramelli, H. Hauffe, and G. Bertorelle. 2003. The Genetic Impact of Demographic Decline and Reintroduction in the Wild Boar (Sus scrofa): A Microsatellite Analysis. *Molecular Ecology* 12:585-595.

Vila, C., and P. Savolainen. 1997. Multiple and Ancient Origins of the Domestic Dog. *Science* 276:1687-1692.

Villa, P., and E. Mahieu. 1991. Breakage Patterns of Human Long Bones. *Journal of Human Evolution* 21:27-48.

Virgos, E. 2002. Factors Affecting Wild Boar (Sus scrofa) Occurrence in Highly Fragmented Mediterranean Landscapes. *Canadian Journal of Zoology* 80:430-435.

von den Driesch, A. 1976. *A Guide to the Measurement of Animal Bones from Archaeological Sites.* Vol. 1. Peabody Museum Bulletins. Cambridge, MA: Peabody Museum, Harvard University.

von den Driesch, A., and J. Boessneck. 1990. "Die Tierreste von der Mykenischen burg Tiryns bei Nauplion/Peloponnes," in *Sonderdruck aus Tiryns Forschungen und Bericht* XI. Edited by K. Kilian, pp. 87-164. Mainz um Rhein: Verlag Philipp von Zabern.

von den Driesch, A., and K. Enderle. 1976. "Die Tierreste aus der Agia Sofia-Magula in Thessalien," in *Magulen um Larisa in Thessalien, 1966, Beitrage zur ur- und fruhgeschichtlichen Archaologie des Mittelmeer-Kulturraumes* 15. Edited by V. Milojcic, A. von den Driesch, K. Enderle, J. Milojcic-von Zumbusch, and K. Kilian, pp. 15-54. Bonn: Habelt.

Wailes, B. Editor. 1996. *Craft Specialization and Social Evolution: In Memory of V.Gordon Childe*. Vol. 6. University Museum Symposium Series, University Museum Monograph 93. Philadelphia: The University Museum of Archaeology and Anthropology, University of Pennsylvania.

Walker, P. L. 1978. Butchering and Stone Tool Function. *American Antiquity* 43:710-715.

Walker, P. L., and J. C. Long. 1977. An Experimental Study of the Morphological Characteristics of Tool Marks. *American Antiquity* 42:605-616.

Wapnish, P., and B. Hesse. 1988. Urbanization and the Organization of Animal Production at Tell Jemmeh in the Middle Bronze Age Levant. *Journal of Near Eastern Studies* 47:81-94.

Warren, P., and V. Hankey. 1989. *Aegean Bronze Age Chronology*. Bristol: Bristol Classical Press.

Watson, J. P. N. 1979. Faunal Remains. *Annual of the British School at Athens* 74:228-229.

Weber, M. 1958. *The City*. Glencoe,Il.: The Free Press.

Weingarten, J., J. H. Crouwel, M. Prent, and G. Vogelsang-Eastwood. 1999. Early Helladic Sealings from Geraki in Lakonia, Greece. *Oxford Journal of Archaeology* 18:357-376.

Weinstock, J. 1993. Two Complete (Sus) Skeletons from Southern Germany: Considerations of Limb Proportions and Ageing Criteria. *Archaeozoologia* 6:71- 92.

White, T. D. 1992. *Prehistoric Cannibalism at Mancos 5MTUMR-2346*. Princeton: Princeton University Press.

Wilkie, N. C., and W. D. E. Coulson. Editors. 1985. *Contributions to Aegean Archaeology: Studies in Honor of William A. McDonald*. Minneapolis: Center for Ancient Studies, University of Minnesota.

Wilson, B., C. Grigson, and S. Payne. Editors. 1982. *Ageing and Sexing Animal Bones from Archaeological Sites*. Vol. 109. British Series. Oxford: BAR.

Wrangham, R., and N. Conklin-Brittain. 2003. Cooking as a Biological Trait. *Comparative Biochemistry and Physiology Part A* 136:35-46.

Wright, H. E. 1985. "Paleoecology, Climatic Change and Aegean Prehistory," in *Contributions to Aegean Archaeology: Studies in Honor of Wiliam A. McDonald*. Edited by N. C. Wilkie and W. D. E. Coulson, pp. 183-195. Minneapolis: Center for Ancient Studies, University of Minnesota.

Wright, H. T. 1978. "Toward an Explanation of the Origin of the State," in *Origins of the State: The Anthropology of Political Evolution.* Edited by R. Cohen and E. R. Service, pp. 49-68. Philadelphia: Institute for the Study of Human Issues.

Wright, J. C. Editor. 2004a. *The Mycenaean Feast.* Princeton: American School of Classical Studies at Athens.

—. 2004b. "Comparative Settlement Patterns During the Bronze Age in the Northeastern Peloponnesos," in *Side-by-Side Survey: Comparative Regional Studies in the Mediterranean World.* Edited by S. E. Alcock and J. F. Cherry, pp. 114-131. Oxford: Oxbow Books.

Yannouli, E. 2003. "Non-domestic Carnivores in Greek Prehistory: A Review," in *Zooarchaeology in Greece: Recent Advances,* British School at Athens Studies 9. Edited by E. Kotjabopoulou, Y. Hamilakis, P. Halstead, C. Gamble, and P. Elefanti, pp. 175-192. London: The Managing Committee, British School at Athens.

Yellen, J. E. 1991. Small Mammals !Kung San Utilization and the Production of Faunal Assemblages. *Journal of Anthropological Archaeology* 10:1-26.

Young, T. C., P. Smith, and P. Mortensen. Editors. 1982. *The Hilly Flanks and Beyond: Essays on the Prehistory of Southwestern Asia. Presented to Robert J. Braidwood November 15, 1982.* Studies in Ancient Oriental Civilization No.36. Chicago: The Oriental Institute of the University of Chicago.

Zeder, M. A. 1991. *Feeding Cities: Specialized Animal Economy in the Ancient Near East.* Washington, D.C.: Smithsonian Institution Press.

—. 1996. "The Role of Pigs in Near Eastern Subsistence: A View from the Southern Levant," in *Retrieving the Past: Essays on Archaeological Research and Methodology in Honor of Gus W. Van Beek.* Edited by J. D. Seger, pp. 297-312. Mississippi: Cobb Institute of Archaeology.

—. 1998. "Pigs and Emergent Complexity in the Ancient Near East," in *Ancestors for the Pigs: Pigs in Prehistory,* vol. 15, MASCA Research Papers in Science and Archaeology. Edited by S. M. Nelson, pp. 109-122. Philadelphia: MASCA.

—. 2003. "Food Provisioning in Urban Societies: A View from Northern Mesopotamia," in *The Social Construction of Ancient Cities.* Edited by M. L. Smith, pp. 156-183. Washington, D.C.: Smithsonian Books.

Zerner, C., P. Zerner, and J. Winder. Editors. 1993. *Wace and Blegen: Pottery as Evidence for Trade in the Aegean Bronze Age 1939-1989.* Amsterdam.

Zeuner, F. E. 1963. *A History of Domesticated Animals.* New York: Harper & Row.

www.ingramcontent.com/pod-product-compliance
Ingram Content Group UK Ltd.
Pitfield, Milton Keynes, MK11 3LW, UK
UKHW061847041026
15591UKWH00015B/947